Basic Concepts of Probability and Evidence in the Law

Michael O. Finkelstein

Basic Concepts
of Probability and Statistics
in the Law

 Springer

Michael O. Finkelstein
25 East 86 St., Apt. 13C
New York NY 10028-0553
USA
mofinkelstein@hotmail.com

ISBN 978-0-387-87500-2   e-ISBN 978-0-387-87501-9
DOI 10.1007/b105519

Library of Congress Control Number: 2008940587

Printed on acid-free paper

springer.com

Dedicated To My Wife, Vivian

Preface

When as a practicing lawyer I published my first article on statistical evidence in 1966, the editors of the *Harvard Law Review* told me that a mathematical equation had never before appeared in the review.[1] This hardly seems possible - but if they meant a *serious* mathematical equation, perhaps they were right. Today all that has changed in legal academia. Whole journals are devoted to scientific methods in law or empirical studies of legal institutions. Much of this work involves statistics. Columbia Law School, where I teach, has a professor of law and epidemiology and other law schools have similar "law and" professorships. Many offer courses on statistics (I teach one) or, more broadly, on law and social science.

The same is true of practice. Where there are data to parse in a litigation, statisticians and other experts using statistical tools now frequently testify. And judges must understand them. In 1993, in its landmark *Daubert* decision, the Supreme Court commanded federal judges to penetrate scientific evidence and find it "reliable" before allowing it in evidence.[2] It is emblematic of the rise of statistics in the law that the evidence at issue in that much-cited case included a series of epidemiological studies. The Supreme Court's new requirement made the Federal Judicial Center's *Reference Manual on Scientific Evidence*, which appeared at about the same time, a best seller. It has several important chapters on statistics.

Before all this began, to meet the need for a textbook, Professor Bruce Levin and I wrote *Statistics for Lawyers*, which was first published in 1990. A second edition appeared in 2000. I use the book in my course, but law students who had not previously been exposed to statistical learning frequently complained that it was too hard. This led me to write a much shorter and mathematically less challenging version - the present book. I have inflicted pamphlet versions of it on students for several years, with good feedback. The book can be read as an introduction to statistics in law standing alone, or in conjunction with other materials.

Where I thought the exposition in *Statistics for Lawyers* not too mathematical to be accessible, I have made free use of it, seeing no reason to restate what we

---

[1] The article is "The Application of Statistical Decision Theory to the Jury Discrimination Cases," 80 Harvard Law Review 338 (1966).

[2] Daubert v. Merrell Dow Pharmaceuticals, Inc., 509 U.S. 579 (1993).

had already written. Not all subjects in *Statistics for Lawyers* are covered; some selectivity was inevitable to keep the book short. On the other hand, a number of new cases are discussed and in some instances I have described in greater detail how the courts have grappled - or not grappled - with statistical evidence. The fate of such evidence in the inevitably messy world of litigation bears only a faint resemblance to the pristine hypotheticals, with their neat conclusions, used in most elementary statistical texts. Among the frustrations of social scientists who work in statistics and law is the fact that interesting statistical questions are not infrequently rendered moot by some point of law or fact that disposes of the case without their resolution, indeed perhaps to avoid their resolution. The book nonetheless considers such cases. The book also includes material from a number of studies that Professor Levin and I made after *Statistics for Lawyers*. I am grateful to him for our long and fruitful collaboration. His current position as Chair of the Department of Biostatistics at Columbia's Mailman School of Public Health has kept him from joining me in writing this text, so any errors that may have crept by me are my responsibility alone.

Finally, I am indebted to my wife, Professor Vivian Berger, for her meticulous reading of certain chapters from the point of view of a legally trained person being introduced to parts of the subject for the first time.

New York, NY   Michael O. Finkelstein
April 28,2008

Contents

Chapter 1
Probability

Classical and Legal Probability

Probability in mathematical statistics is classically defined in terms of the outcomes of conceptual experiments, such as tossing ideal coins and throwing ideal dice. In such experiments the probability of an event, such as tossing heads with a coin, is defined as its relative frequency in long-run trials. Since the long-run relative frequency of heads in tosses of a fair coin "closes in" on one-half, we say that the probability of heads on a single toss is one-half. Or, to take a more complicated example, if we tossed a coin 50 times and repeated the series many times, we would tend to see 30 or more heads in 50 tosses only about 10% of the time; so we say that the probability of such a result is one-tenth. We refer to this relative frequency interpretation as classical probability. Calculations of classical probability generally are made assuming the underlying conditions by which the experiment is conducted, in the above examples with a fair coin and fair tosses.

This is not to say that the ratio of heads in a reasonably large number of tosses invariably equals the probability of heads on a single toss. Contrary to what some people think, a run of heads does not make tails more likely to balance out the results. Nature is not so obliging. All she gives us is a fuzzier determinism, which we call the law of large numbers. It was originally formulated by Jacob Bernoulli (1654–1705), the "bilious and melancholy" elder brother of the famous Bernoulli clan of Swiss mathematicians, who was the first to publish mathematical formulas for computing the probabilities of outcomes in trials like coin tosses. The law of large numbers is a formal statement, proved mathematically, of the vague notion that, as Bernoulli biliously put it, "Even the most stupid of men, by some instinct of nature, by himself and without any instruction (which is a remarkable thing), is convinced that the more observations have been made, the less danger there is in wandering from one's goal."[1]

---

[1] J. Bernoulli, *Ars Conjectandi* [The Art of Conjecturing] 225 (1713), quoted in S.M. Stigler, *The History of Statistics: The Measurement of Uncertainty Before 1900* at 65 (1986).

M.O. Finkelstein, *Basic Concepts of Probability and Statistics in the Law*,
DOI 10.1007/b105519_1, © Springer Science+Business Media, LLC 2009

To understand the formal content of the commonplace intuition, think of the difference between the ratio of successes in a series of trials and the probability of success on a single trial as the error of estimating the probability from the series. Bernoulli proved that the probability that the error exceeds any given arbitrary amount can be made as small as one chooses by increasing sufficiently the number of trials; it is in this sense that the long-run frequency "closes in" on the probability in a single trial. The law of large numbers represented a fateful first step in the process of measuring the uncertainty of what has been learned from nature by observation. Its message is obvious: The more data the better.

What has classical probability to do with the law? The concept of probability as relative frequency is the one used by most experts who testify to scientific matters in judicial proceedings. When a scientific expert witness testifies that in a study of smokers and non-smokers the rate of colon cancer among smokers is higher than the rate among non-smokers and that the difference is statistically significant at the 5% level, he is making a statement about long-range relative frequency. What he means is that *if* smoking did not cause colon cancer and *if* repeated samples of smokers and non-smokers were drawn from the population to test that hypothesis, a difference in colon cancer rates at least as large as that observed would appear less than 5% of the time. The concept of statistical significance, which plays a fundamental role in science, thus rests on probability as relative frequency in repeated sampling.

Notice that the expert in the above example is addressing the probability of the data (rates of colon cancer in smokers and non-smokers) given an hypothesis about the cause of cancer (smoking does not cause colon cancer). However, in most legal settings, the ultimate issue is the inverse conditional of that, i.e., the probability of the cause (smoking does not cause colon cancer) given the data. Probabilities of causes given data are called *inverse probabilities* and in general are not the same as probabilities of data given causes. In an example attributed to John Maynard Keynes, if the Archbishop of Canterbury were playing poker, the probability that the Archbishop would deal himself a straight flush given honest play on his part is not the same as the probability of honest play on his part given that he has dealt himself a straight flush. The first is 36 in 2,598,960; the second most people would put at close to 1 (he is, after all, an archbishop).

One might object that since plaintiff has the burden of proof in a law suit, the question in the legal setting is not whether smoking does not cause cancer, but whether it does. This is true, but does not affect the point being made here. The probability that, given the data, smoking causes colon cancer is equal to one minus the probability that it doesn't, and neither will in general be equal to the probability of the data, assuming that smoking doesn't cause colon cancer.

The inverse mode of probabilistic reasoning is usually traced to Thomas Bayes, an English Nonconformist minister from Tunbridge Wells, who was also an amateur mathematician. When Bayes died in 1761 he left his papers to another minister, Richard Price. Although Bayes evidently did not know Price very well there was a good reason for the bequest: Price was a prominent writer on mathematical subjects and Bayes had a mathematical insight to deliver to posterity that he had withheld during his lifetime.

Among Bayes's papers Price found a curious and difficult essay that he later entitled, "Toward solving a problem in the doctrine of chances." The problem the essay addressed was succinctly stated: "*Given* the number of times in which an unknown event has happened and [has] failed: *Required* the chance that the probability of its happening in a single trial lies somewhere between any two degrees of probability that can be named." Price added to the essay, read it to the Royal Society of London in 1763, and published it in *Philosophical Transactions* in 1764. Despite this exposure and the independent exploration of inverse probability by Laplace in 1773, for over a century Bayes's essay remained obscure. In fact it was not until the twentieth century that the epochal nature of his work was widely recognized. Today, Bayes is seen as the father of a controversial branch of modern statistics eponymously known as Bayesian inference and the probabilities of causes he described are called Bayesian or inverse probabilities.

Legal probabilities are mostly Bayesian (i.e., inverse). The more-likely-than-not standard of probability for civil cases and beyond-a-reasonable-doubt standard for criminal cases import Bayesian probabilities because they express the probabilities of events given the evidence, rather than the probabilities of the evidence, given events. Similarly, the definition of "relevant evidence" in Rule 401 of the Federal Rules of Evidence is "evidence having any tendency to make the existence of any fact that is of consequence to the determination of the action more probable or less probable than it would be without the evidence." This definition imports Bayesian probability because it assumes that relevant facts have probabilities attached to them. By contrast, the traditional scientific definition of relevant evidence, using classical probability, would be any "evidence that is more likely to appear if any fact of consequence to the determination of the action existed than if it didn't."

The fact that classical and Bayesian probabilities are different has caused some confusion in the law. For example, in an old case, *People v. Risley*,[2] a lawyer was accused of removing a document from the court file and inserting a typed phrase that helped his case. Eleven defects in the typewritten letters of the phrase were similar to those produced by defendant's machine. The prosecution called a professor of mathematics to testify to the chances of a random typewriter producing the defects found in the added words. The expert assumed that each defect had a certain probability of appearing and multiplied these probabilities together to come up with a probability of one in four billion, which he described as "the probability of these defects being reproduced by the work of a typewriting machine, other than the machine of the defendant. . .." The lawyer was convicted. On appeal, the New York Court of Appeals reversed, expressing the view that probabilistic evidence relates only to future events, not the past. "The fact to be established in this case was not the probability of a future event, but whether an occurrence asserted by the People to have happened had actually taken place."[3]

---

[2] 214 N.Y. 75 (1915).

[3] *Id.* at 85.

There are two problems with this objection. First, the expert did not compute the probability that defendant's machine did not type the insert. Although his statement is somewhat ambiguous, he could reasonably be understood to refer to the probability that there would have been matching defects *if* another machine had been used. Second, even if the expert had computed the probability that the insert had been typed on defendant's machine, the law, as we have seen, *does* treat past events as having probabilities.[4] If probabilities of past events are properly used to define the certainty needed for the final verdict, there would seem to be no reason why they are not properly used for subsidiary issues leading up to the final verdict. As we shall see, the objection is not to such probabilities per se, but to the expert's competence to calculate them.

A similar confusion arose in a notorious case in Atlanta, Georgia. After a series of murders of young black males, one Wayne Williams was arrested and charged with two of the murders. Critical evidence against him included certain unusual trilobal fibers found on the bodies. These fibers matched those in a carpet in Williams' home. A prosecution expert testified that he estimated that only 82 out of 638,992 occupied homes in Atlanta, or about 1 in 8,000, had carpeting with that fiber. This type of statistic has been called "population frequency" evidence. Based on this testimony, the prosecutor argued in summation that "there would be only one chance in 8,000 that there would be another house in Atlanta that would have the same kind of carpeting as the Williams home." On appeal, the Georgia Court of Appeals rejected a challenge to this argument, holding that the prosecution was not precluded from "suggesting inferences to be drawn from the probabilistic evidence."

Taken literally, the prosecutor's statement is nonsense because his own expert derived the frequency of such carpets by estimating that 82 Atlanta homes had them. To give the prosecutor the benefit of the doubt, he probably meant that there was one chance in 8,000 that the fibers came from a home other than the defendant's. The 1-in-8,000 figure, however, is not that, but the probability an Atlanta home picked at random would have that fiber.

Mistakes of this sort are known as the fallacy of the inverted conditional. That they should occur is not surprising. It is not obvious how classical probability based, for example, on population frequency evidence bears on the probability of defendant's criminal or civil responsibility, whereas Bayesian probability purports to address the issue directly. In classical terms we are given the probability of seeing the incriminating trace if defendant did not leave it, but what we really want to know is the probability that he did leave it. In a litigation, the temptation to restate things in Bayesian terms is very strong. The Minnesota Supreme Court was so impressed by the risk of this kind of mistake by jurors that it ruled out population frequency evidence, even when correctly stated.[5] The court apprehended a "real danger that the jury will use the evidence as a measure of the probability of the defendant's

---

[4] To be sure, the court's rejection of the testimony was correct because there were other, valid objections to it. See p. 49.

[5] State v. Kim, 398 N.W. 2d 544 (1987); State v. Boyd, 331 N.W. 2d 480 (1983).

guilt or innocence."[6] The court evidently feared that if, for example, the population frequency of a certain incriminating trace is 1 in a 1,000 the jury might interpret this figure as meaning that the probability of defendant's innocence was 1 in a 1,000. And, as we have seen, it is not only jurors who can make such mistakes. This particular misinterpretation, which arises from inverting the conditional, is sometimes called the prosecutor's fallacy.[7]

Is the prosecutor's fallacy in fact prejudicial to the accused? A study using simulated cases before juries of law students showed higher rates of conviction when "prosecutors" were allowed to misinterpret population frequency statistics as probabilities of guilt than when the correct statement was made.[8] The effect was most pronounced when population frequencies were as high as one in a thousand, but some effect also appeared for frequencies as low as one in a billion. The study suggests that the correctness of interpretation may matter.

The defendant also has his fallacy, albeit of a different sort. This is the argument that the evidence does no more than put defendant in a group consisting of all those who have the trace in question, so that the probability that defendant left the trace is only one over the size of the group. If this were correct, then only a show of uniqueness (which is perhaps possible for DNA evidence, but all but impossible as a general matter) would permit us to identify a defendant from a trace. This is not a fallacy of the inverted conditional, but is fallacious because, as we shall see, it ignores the other evidence in the case.

Bayes's Theorem

To examine the prosecutor's and defendant's fallacies a little more closely I ask a more general question: If it is wrong to interpret the probabilities of evidence given assumed causes as probabilities of causes given assumed evidence, what is the relation between the two probabilities? Specifically, what is the probative significance of scientific probabilities of the kind generated by statistical evidence to the probabilities of causes implied by legal standards? The answer is given by what is now called Bayes's theorem, which Bayes derived for a special case using a conceptual model involving billiard balls. We do not give his answer here. Instead, to explain what his result implies for law, we use a doctored example

---

[6] State v. Boyd, 331 N.W. 2d at 483.

[7] The Minnesota Supreme Court's opinions seem unduly restrictive of valuable evidence and its rulings were overturned by statute with respect to genetic or blood test results. The statute provides: "In a civil or criminal trial or hearing, statistical population frequency evidence, based on genetic or blood test results, is admissible to demonstrate the fraction of the population that would have the same combination of genetic markers as was found in a specific human biological specimen. 'Genetic marker' means the various blood types or DNA types that an individual may possess." Minn. Stat. Sec. 634.26 (2008).

[8] D.L. Faigan & A.J. Baglioni, Jr., *Bayes' Theorem in the Trial Process*, 12 Law and Human Relations 1 (1988).

from a law professor's chestnut: the case of the unidentified bus.[9] It is at once more general and mathematically more tractable than the problem that Bayes addressed.

The facts are simply stated. On a rainy night, a driver is forced into collision with a parked car by an unidentified bus. Of the two companies that run buses on the street, Company A owns 85% of the buses and Company B owns 15%. Which company was responsible? That icon of the law, an eyewitness, testifies that it was a Company B bus. A psychologist testifies without dispute that eyewitnesses in such circumstances tend to be no more than 80% accurate. Our problem is to find the probabilities associated with the cause of the accident (Companies A or B) given the case-specific evidence (the eyewitness report) and the background evidence (the market shares of Companies A and B). To be specific, let us ask for the probability that it was a Company B bus, assuming that the guilty bus had to belong to either Company A or Company B.

The Bayesian result of relevance to this problem is most simply stated in terms of odds:[10] The posterior odds that it was a Company B bus are equal to the prior odds that it was a Company B bus times the likelihood ratio that it was such a bus. Thus,

$$posterior\ odds = prior\ odds \times likelihood\ ratio.$$

In this formula the *posterior odds* are the odds that the cause of the accident was a Company B bus, given (or posterior to) the background and case-specific evidence. These are Bayesian odds. The *prior odds* are the odds that the cause of the accident was a Company B bus prior to considering the case-specific evidence. These are also Bayesian. If one assumes in our bus problem that, in the absence of other evidence bearing on routes and schedules, the prior probabilities are proportional to the sizes of the respective bus fleets, the probability that it was a Company A bus is 0.85 and a Company B bus is 0.15. Hence the prior odds that the bus was from Company A are $0.85/0.15 = 5.67$ and from Company B are $0.15/0.85 = 0.1765$. Since probabilities greater than 50% and odds greater than 1 meet the more-likely-than-not standard for civil cases, plaintiff should have enough for a verdict against Company A, *if* we allow the sufficiency of statistical evidence. That is a big "if." I return to this point later.

To solve the problem that he set for himself, Bayes made the restrictive assumption that the prior probabilities for the causes were equal. Prior probability distributions that assign equal or nearly equal probabilities to the possible causes are now known as diffuse or flat priors because they do not significantly favor one possibility

---

[9] The chestnut is from Smith v. Rapid Transit, Inc., 307 Mass. 469 (1945). The doctoring is from D. Kahneman, A. Tversky & P. Slovic (eds.), *Judgment Under Uncertainty: Heuristics and Biases* 1 (1982).

[10] The odds on an event are defined as the probability that the event will occur, p, divided by the probability that it will not, $1 - p$. Conversely, the probability of an event is equal to the odds on the event divided by one plus the odds.

over another. The prior in our example is informative and not diffuse or flat because it assigns much greater probability to one possible cause than to the other.

The third element in Bayes's theorem is the *likelihood ratio* for the event given the evidence. The likelihood ratio for an event is defined as the probability of the evidence if the event occurred divided by the probability of the evidence if the event did not occur. These are classical probabilities because they are probabilities of data given causes. In our bus example, the likelihood ratio that it was a Company B bus given the eyewitness identification is the probability of the witness reporting that it was a Company B bus if in fact it was, divided by the probability of such a report if it were not. On the facts stated, the numerator is 0.80, since the witness is 80% likely to report a Company B bus if it was such a bus. The denominator is 0.20, since 20% of the time the witness would mistakenly report a Company B bus when it was a Company A bus. The ratio of the two is $0.80/0.20 = 4$. We are four times as likely to receive a report that it was a Company B bus if it was in fact such a bus than if it was not.

The likelihood ratio is an important statistical measure of the weight of evidence. It is intuitively reasonable. The bloody knife found in the suspect's home is potent evidence because we think we are far more likely to find such evidence if the suspect committed the crime than if he didn't. In general, large values of the likelihood ratio imply that the evidence is strong, and small values greater than 1 that it is weak. A ratio of 1 means that the evidence has no probative value; and values less than 1 imply that the evidence is exonerating.

Putting together the prior odds and the likelihood ratio, the posterior odds that it was a Company B bus given the evidence are $0.1765 \times 4.00 = 0.706$. The probability that it was a Company B bus is $0.706/(1 + 0.706) = 0.4138$. Thus, despite eyewitness identification, the probability of a Company B bus is less than 50%. If there were a second eyewitness with the same testimony and the same accuracy, the posterior odds with respect to the first witness could be used as the prior odds with respect to the second witness and Bayes's theorem applied again. In that case the new posterior odds would be $0.706 \times 4.00 = 2.824$ and the new posterior probability would be $2.824/3.824 = 0.739$.

Some people object to these results. They argue that if the eyewitness is right 80% of the time and she says it was a Company B bus, why isn't the probability 80% that it was a Company B bus? Yet we find that, despite the eyewitness, the preponderance of probability is against the witness's testimony. The matter is perhaps even more counterintuitive when there are two eyewitnesses. Most people would think that two eyewitnesses establish a proposition beyond a reasonable doubt, yet we conclude that the probability of their being correct is only about 74%, even when there is no contrary testimony. Surely Bayes's theorem is off the mark here.

But this argument confuses two conditional probabilities: the probability that the witness would so testify conditional on the fact that it was a Company B bus (which is indeed 80%) and the probability that it was a Company B bus conditional on the fact that the witness has so testified (which is not necessarily 80%; remember the Archbishop playing poker).

The second and related answer to the objection to the Bayesian result is that the 80% figure ignores the effect of the statistical background, i.e., the fact that there are many more Company A buses than Company B buses. For every 100 buses that come along only 15 will be Company B buses but 17 (0.20 × 85) will be Company A buses that are wrongly identified by the first witness. Because of the fact that there are many more Company A buses, the witness has a greater chance of wrongly identifying a Company A bus as a Company B bus than of correctly identifying a Company B bus. The probabilities generated by Bayes's theorem reflect that fact. In this context its application corrects for the tendency to undervalue the evidentiary force of the statistical background in appraising the case-specific evidence.[11] This appears to be a general phenomenon.[12]

Screening Tests

The correction for statistical background supplied by Bayes's theorem becomes increasingly important when the events recorded in the background are rare. In that case even highly accurate particular evidence may become surprisingly inaccurate. Screening devices are of this character. The accuracy of such screening tests is usually measured by their sensitivity and specificity. *Sensitivity* is the probability that the test will register a positive result if the person has the condition for which the test is given. *Specificity* is the probability the test will be negative if the person doesn't have the condition. Together, these are the test's *operating characteristics.*

For example, the Federal Aviation Administration is said to use a statistically based hijacker profile program to help identify persons who might attempt to hijack a plane using a nonmetallic weapon. Assume that the test has a sensitivity of 90% (i.e., 90% of all hijackers are detected) and a specificity of 99.95% (i.e., 99.95% of all non-hijackers are correctly identified). This seems and is very accurate. But if the rate of hijackers is 1 in 25,000 passengers, Bayes's theorem tells us that this seemingly accurate instrument makes many false accusations.

The odds that a passenger being identified as a hijacker by the test is actually a hijacker are equal to the prior odds of a person being a hijacker times the likelihood ratio associated with the test. Our assumption is that the prior odds that a passenger is a hijacker are $(1/25,000)/(24,999/25,000) = 1/24,999$). The likelihood ratio for the test is the probability of a positive identification if the person is a hijacker (0.90) divided by the probability of such an identification if the person is not $(1 - 0.9995 = 0.0005)$. The ratio is thus $0.90/0.0005 = 1,800$. The test is powerful evidence, but hijackers are so rare that it becomes quite inaccurate. The posterior odds of a correct

---

[11] Empirical studies based on simulated trials show a similar underweighting. See, e.g., J. Goodman, "Jurors' Comprehension and Assessment of Probabilistic Evidence", 16 American Journal of Trial Advocacy 361 (1992).

[12] See, e.g., *Judgment Under Uncertainty: Heuristics and Biases* at 4–5 (D. Kahneman, P. Slovic, and A. Tversky, eds., 1982).

identification are only $(1/24{,}999) \times 1{,}800 = 0.072$. The posterior probability of a correct identification is only $0.072/1.072 = 0.067$; there is only a 6.7% chance that a person identified as a hijacker by this accurate test is really a hijacker. This result has an obvious bearing on whether the test affords either probable cause to justify an arrest or even reasonable suspicion to justify a brief investigative detention.[13]

The failure to distinguish posterior odds and likelihood ratios has led to some confusion in the Supreme Court's jurisprudence on the constitutionality of stopping passengers to search for drugs based on a profile.[14] In particular, debates have swirled around the question whether a profile can be a basis for "reasonable suspicion" that the person is a drug courier, which would justify a brief investigative detention, a lower standard than "probable cause" needed for an arrest. Since drug couriers are rare among passengers, even profiles with impressive operating characteristics are likely to generate small posterior odds that the person is a courier. For example, imagine a profile that is quite accurate: 70% of drug couriers would fit the profile, but only 1 normal passenger in a 1,000 would do so. The likelihood ratio for the profile is $0.70/0.001 = 700$, which means that the odds that a person is a courier are increased 700-fold by matching the profile. But if the rate of couriers is 1 in 10,000 passengers, the odds on the person matching the profile is a courier are only $(1/9{,}999) \times 700 = 0.07$, or a probability of 6.5%. This result is not fanciful: A study by the U.S. Customs Service found that the hit rate for such stops was about 4%.

These facts lead us to the legal issue that the Supreme Court has not squarely faced in these cases: whether a profile that significantly increases the probability that the person is a criminal is a sufficient basis for reasonable suspicion, even if that increased probability remains low because couriers are so rare. By greatly increasing the probability that a person who fits the profile is a courier, a valid profile clearly can provide a rational basis for investigation. If "reasonable suspicion" requires no more than a rational basis for suspicion, then a profile could be sufficient. On the other hand, if "reasonable suspicion" requires a reasonable probability of crime, a profile is unlikely to be sufficient. Which standard applies is presently unclear.

We have been considering the effects of the likelihood ratios of screening tests on their predictive values. We now turn things around and consider the implications of predictive values for likelihood ratios. In *United States v. Scheffer*,[15] the Supreme Court considered the constitutionality of Military Rule of Evidence 707, which made polygraph evidence inadmissible in military courts-martial. Scheffer challenged the rule because it denied him the right to introduce an exonerating test. Justice Thomas, writing for the majority, upheld the rule, arguing that the reliability of the polygraph was uncertain; he noted that views of polygraph accuracy ranged from 87% to little better than the coin toss (50%). However, in taking the posi-

---

[13] *Cf.* United States v. Lopez, 328 F. Supp. 1077 (E.D.N.Y. 1971).

[14] For a discussion of the Supreme Court cases, see M.O. Finkelstein & B. Levin, "On the Probative Value of Evidence from a Screening Search," 43 Jurimetrics Journal 265, 270–276 (2003).

[15] 523 U.S. 303 (1998).

tion that polygraphs were not sufficiently reliable to be admitted in evidence, the government had to deal with the interesting fact that the Defense Department itself used polygraph tests to screen employees in its counterintelligence program. If good enough for counterintelligence, why not for court?

The negative predictive values of the test were very strong in the counterintelligence context: If an employee passed the test the probabilities were very high that he was not a security risk meriting further investigation. But this high value is probably not applicable to polygraphs in criminal cases because it depends on the operating characteristics of the test *and* on the prior odds of guilt; and the prior odds on guilt would probably be much greater in a criminal case than in a screening program. Nevertheless, admissibility should not turn on the test's predictive values because the strength of other evidence of guilt is no justification for keeping out strong exonerating test results. Probative strength for these purposes is more appropriately measured by the likelihood ratio associated with a negative test result, which measures how much the predictive values change with the results. Information about the likelihood ratio can be gleaned from the results of screening, assuming that the likelihood ratio remains the same across contexts. (This is a conservative assumption because it is probable that test is more accurate in the criminal context than in the more amorphous arena of security screening.)

In 1997, about the time of *Scheffer*, the Defense Department used the polygraph to screen 7,616 employees; the polygraph identified 176 employees as possible security risks, to be further investigated, and passed 7,440 employees. Of the 176 employees at the time the data were reported to Congress, 6 had lost clearance and the cases of 16 were pending. Assume for the moment that all pending cases were resolved unfavorably to the individuals, so there would be 22 security risks among the 176 identified by the polygraph screen. How many security risks were not identified by the polygraph? If the polygraph has only a 50% chance of identifying a security risk (Justice Thomas's coin toss), there would be another 22 risks that were not so identified; the negative predictive value of the test would be 7,418/7,440 = 0.997, and the negative predictive odds would be 0.997/0.003 = 332. Since there are assumed to be 44 security risks out of 7,616 employees, the prior probability that an employee is not a security risk is 7,572/7,616 = 0.994, and the prior negative odds are 0.994/0.006 = 166. Using Bayes's theorem, the likelihood ratio associated with a negative polygraph test is therefore about 332/166 = 2.0. If the polygraph were 87% accurate, a similar calculation shows that the likelihood ratio for an exonerating test would rise to about 8. Thus the odds on innocence are increased between two and eight times by an exonerating polygraph.[16] An incriminating polygraph has low predictive power in the counterintelligence arena – between 3.4 and 12.5% – but an even stronger likelihood ratio than the exonerating polygraph. The odds on guilt are increased by between 25 and 43 times for a defendant who fails a polygraph test. Looking at even the weakest result (the likelihood ratio not quite 2), most lawyers

---

[16] These results are not much changed if it is assumed that none of the pending cases resulted in adverse action against individuals. In that scenario, the likelihood ratio is between 1.95 and 6.8, respectively.

would regard evidence that almost doubled the odds on innocence to be legally significant and certainly sufficiently probative to be admissible.[17] The evidentiary value of either outcome, therefore, seems enough to make the case that the lack of reliability should not be a ground for exclusion of test results, as either incriminating or exonerating evidence.(There may, of course, be other reasons, as Justice Thomas argued, for excluding such evidence.[18]) The larger message here is that a test that is used for screening purposes, even one with low positive predictive values, is likely to have sufficient reliability to be admissible in judicial proceedings.

Debate over Bayesian analysis

Classical statisticians have two basic objections to Bayesian analysis. The first is philosophical. They point out that for events above the atomic level, which is the arena for legal disputes, the state of nature is not probabilistic; only the inferences to be drawn from our evidence are uncertain. From a classical statistician's point of view, Bayesians misplace the uncertainty by treating the state of nature itself, rather than our measurement of it, as having a probability distribution. This was the point of view of the *Risley* court. However, the classicists are not completely consistent in this and sometimes compute Bayesian probabilities when there are data on which to base prior probabilities.

The second objection is more practical. In most real-life situations, prior probabilities that are the starting point for Bayesian calculations cannot be based on objective quantitative data of the type at least theoretically available in our examples, but can only reflect the strength of belief in the proposition asserted. Such probabilities are called subjective or personal probabilities. They are defined not in terms of relative frequency, but as the odds we would require to bet on the proposition. While the late Professor Savage has shown that subjective or personal probabilities satisfying a few seemingly fundamental axioms can be manipulated with the same methods of calculation used for probabilities associated with idealized coins or dice, and thus are validly used as starting points in Bayesian calculations, there is a sharp dispute in the statistical community over the acceptability of numerical measures of persuasion in scientific calculations.[19] After all, subjective prior probabilities may vary

---

[17] The definition of "relevant evidence" in the Federal Rules of Evidence is "evidence having any tendency to make the existence of any fact that is of consequence to the determination of the action more probable or less probable than it would be without the evidence." FED. R. EVID. 401. A polygraph test with an LR of about 2 is certainly relevant evidence by that definition.

[18] *Scheffer*, 523 U.S. at 312–17. Justice Thomas gave as additional reasons for sustaining Rule 707 that the exclusion of polygraph evidence preserved the court members' core function of making credibility determinations, avoided litigation over collateral issues, and did not preclude the defense from introducing any factual evidence, as opposed to expert opinion testimony.

[19] Whether people conform their personal probabilities to the assumptions has also been questioned.

from person to person without a rational basis. No other branch of science depends for its calculations on such overtly personal and subjective determinants.

To these objections the Bayesians reply that since we have to make Bayesian judgments in any event, at some point we must introduce the very subjectivity that we reject in the classical theory. This answer seems particularly apt in law. We require proof that makes us believe in the existence of past events to certain levels of probability; this attachment of probabilities to competing states of nature is fundamentally Bayesian in its conception. And since we have to make these judgments in any event, the Bayesians argue that it is better to bring them within the formal theory because we can correct for biases that are far more important than the variability of subjective judgments with which we started.

To make this point specific, let me report on my personal experiments with the theory. I give each class of my law students the following hypothetical and ask them to report their personal probabilities.[20] A woman is found in a ditch in an urban area. She has been stabbed with a knife found at the scene. The chief suspect is her boyfriend. They were together the day before and were known to have quarreled. He once gave her a black eye. Based on these facts, I ask the students for their estimate of the probability that he killed her. For the most part the students give me estimates between 25 and 75%. Now I tell them that there is a partial palm print on the knife, so configured that it is clear that it was left there by someone who used it to stab rather than cut. The palm print matches the boyfriend's palm, but it also appears in one person in a thousand in the general population. I ask again, what is the probability that the killed her, assuming that the test has no error and if it is his print then he is guilty. Usually (not always!) the estimate goes up, but generally to less than 90%.

Bayes's theorem teaches us that the adjustments made by the students are too small. For the person whose prior was 0.25, his or her posterior probability should be 0.997.[21] At the upper end of the range, the prior probability is 0.75 and the posterior probability should be 0.9997. The difference between these probabilities is not significant for decision purposes: We convict in either case. The point is that underestimation of the force of statistical evidence (the 1 in a 1,000 statistic) when it is informally integrated with other evidence is a source of systematic bias that is far more important than the variation due to subjectivity in estimating the prior probability of guilt, even assuming that the subjective variation is entirely due to error.

Besides correcting for bias, Bayes's theorem helps shed some light on the value of traces used for identification. Take the prosecutor's and defendant's fallacies pre-

---

[20] Taken from M.O. Finkelstein & W. Fairley, "A Bayesian Approach to Identification Evidence," 83 Harvard Law Review 489 (1970).

[21] For such a person the prior odds are $0.25/0.75 = 1/3$. The likelihood ratio is 1 (we are certain that the palm print would be like the defendant's if he left it) divided by $1/1,000$ (the rate of such palm prints in the population if he didn't leave it), or 1,000. Applying Bayes's theorem, $1/3 \times 1,000 = 333.33$ (the posterior odds). The posterior probability of guilt is thus $333.33/334.33 = 0.997$.

viously discussed. We see from Bayes's theorem that these two arguments turn on the assumed prior probabilities. If the prior probability that the accused was the source of the trace is 50%, then the prior odds are $0.50/0.50 = 1$ and the posterior odds given the match are equal to the likelihood ratio, which is the reciprocal of the population frequency. Thus if the population frequency of an observed trace is 1 in 1,000, the posterior odds that the accused left it are 1,000 and the posterior probability that the accused left the print is $1,000/1,001 = 0.999$, which is as the prosecutor asserted. The evidence is so powerful that it could not be excluded on the ground that the jury might overvalue it. On the other hand, if there is no evidence of responsibility apart from the trace, the accused is no more likely to have left it than anyone else in the relevant population who shared the trace. If N is the size of the relevant population (including the accused) then there are $N/1,000$ people in the population who share the trace. So while the prosecutor's fallacy assumes a 50% prior probability that the accused left the print, the defendant's fallacy assumes a prior probability no greater than that of a randomly selected person from those possessing the trace.

Neither assumption is likely to be justified in a real criminal case in which there is (as there will almost certainly be) other evidence of guilt, but given such other evidence the prosecutor's fallacy is likely to be closer to the truth than the defendant's fallacy.

The prosecutor's fallacy was on full display in the U.K. case of *Regina v. Alan Doheny & Gary Adams*, which involved two appeals from convictions for sexual assault. The Crown's experts testified that the "random occurrence ratios" for matching DNA stains taken from crime semen and the defendants were 1 in 40 million in *Doheny* and 1 in 27 million in *Adams*.[22] In both cases the Crown's expert testified, in substance, that the chances of the semen having come from someone other than the accused were less than the random occurrence ratio. In its introduction to its separate opinions in the two cases, the court found that such testimony reflected the prosecutor's fallacy, the falsity of which it demonstrated with the following example: "If one person in a million has a DNA profile which matches that obtained from the crime stain, then the suspect will be one of perhaps 26 men in the United Kingdom who share that characteristic. If no fact is known about the Defendant other than that he was in the United Kingdom at the time of the crime the DNA evidence tells us no more that there is a statistical probability that he was the criminal of 1 in 26." In the court's view, the DNA expert should confine himself to giving the random occurrence ratio and perhaps the number of persons in the United Kingdom (or a subarea) with that profile and should not give or suggest a probability that the defendant left the trace.

The court, however, did not quash either conviction on the ground of the prosecutor's fallacy. In the court's view, the fallacy did not "alter the picture" if the random occurrence ratio was as extreme as claimed by the Crown's experts. But this can be true, as we have seen, only if there is enough other evidence linking the accused to

---

[22] Regina v. Alan James Doheny and Gary Adams, [1996] EWCA Crim. 728 (July 31, 1996).

the crime so that the prior probability of the accused's guilt approaches 50% and in any event it is not for the expert to assume that such is the case. The court quashed Doheny's conviction and affirmed Adams's on other grounds.

There is one thread in legal thought that is not consistent with Bayesian probability and that is the attitude toward "naked" or "bare" statistical evidence, i.e., statistical evidence standing alone, without case-specific evidence. The absence of case-specific evidence should generally mean (there are exceptions, to be discussed) that the likelihood ratio is 1 and the posterior odds are equal to the prior odds – which are defined by the statistics. But a bare statistical case is usually said to be ipso facto insufficient, even when there are strong posterior odds that would be sufficient if there were case-specific evidence. The origins of this idea go back to some early cases in which judges, in dicta, sounded the theme that quantitative probability is not evidence. For example, in *Day v. Boston & Maine R.R.*,[23] a case involving the need to determine the cause of death in a railway accident, but in which there was really no evidence of how the accident had occurred, Judge Emery gratuitously commented that mathematical probability was akin to pure speculation:

> Quantitative probability . . . is only the greater chance. It is not proof, nor even probative evidence, of the proposition to be proved. That in one throw of dice there is a quantitative probability, or greater chance, that a less number of spots than sixes will fall uppermost [the probability is 35/36] is no evidence whatever, that in a given throw such was the actual result . . . . The slightest real evidence that sixes did in fact fall uppermost would out weigh all the probability otherwise.[24]

This theme was picked up in *Sargent v. Massachusetts Accident Company*,[25] in which the court had to decide whether the deceased, canoeing in the wilderness, met his end by accident (which would have been within the insurance policy) or by other causes such as starvation (which would not). Of course, there were no data. Citing *Day*, Justice Lummus elaborated, in much-quoted language, the theme of mathematical probability as insufficient proof:

> It has been held not enough that mathematically the chances somewhat favor a proposition to be proved; for example, the fact that colored automobiles made in the current year outnumber black ones would not warrant a finding that an undescribed automobile of the current year is colored and not black, nor would the fact that only a minority of men die of cancer warrant a finding that a particular man did not die of cancer. The weight or preponderance of the evidence is its power to convince the tribunal which has the determination of the fact, of the actual truth of the proposition to be proved. After the evidence has been weighed, that proposition is proved by a preponderance of the evidence if it is made to appear more likely or probable in the sense that actual belief in its truth, derived from the evidence, exists in the mind or minds of the tribunal notwithstanding any doubts that may linger there.[26]

---

[23] 96 Me. 207 (1902).

[24] *Id.* at 217–218.

[25] 307 Mass. 246 (1940).

[26] 307 Mass. at 250 (citations omitted).

It was this language that was cited by the court in *Smith v. Rapid Transit, Inc.*, a case in which there was at least a market-share datum.

Since the *Sargent* dictum cited in *Smith* originated in cases in which there were case-specific facts that were only suggestive, but not compelling, the notion that probability is not evidence seems to spring from the false idea that the uncertainty associated with mathematical probability is no better than ungrounded speculation to fill in gaps in proof. There is of course a world of difference between the two. As Bayes's theorem shows us, the former, when supported, would justify adjusting our view of the probative force of particular evidence, while the latter would not. Nor is it correct, as we have seen, that the slightest real evidence (presumably an eyewitness would qualify) should outweigh all the probability otherwise.

Yet the view that "bare" statistics are insufficient persists.[27] Discussing *Smith*, Judge Posner argued that plaintiff needs an incentive to ferret out case-specific evidence, in addition to the statistics, so he must lose unless he shows that it was infeasible for him to get more particular evidence. Unless he makes that showing, his failure to produce more evidence must be due to the fact either that the other evidence was adverse or that he failed to make an investigation – and in either case he should lose. Judge Posner also objects that if statistics were sufficient, Company B (the Company with fewer buses) would enjoy immunity from suit, all the errors being made against Company A, a significant economic advantage.[28] In the examples cited by Judge Posner, the insufficiency of bare statistical evidence is not necessarily inconsistent with Bayesian formulations. In particular, where the case-specific evidence is not produced because it is, or may be, adverse, the likelihood ratio associated with its absence would be less than 1, weakening the posterior odds to that extent. But the cases rejecting statistical evidence have not been so confined.

An extreme recent case, *Krim v. pcOrder.com, Inc.*,[29] shows that the ghosts of *Smith* and *Sargent*, assisted by Judge Posner, still walk in the law. In that case, the issue was whether the plaintiffs had standing to sue for fraud in the sale of securities under the federal Securities Act of 1933. Since the act was designed to protect only purchasers in public offerings by the company or its underwriters, the court held that plaintiffs had to trace at least one of their shares to those the company issued in the public offering. There were approximately 2.5 million shares in the pool of shares so issued, but when two of the plaintiffs bought their shares from this pool there were intermingled with it shares held by insiders that were also offered for sale. The Securities Act did not apply to those insider shares. Despite the fact that the pool consisted overwhelmingly of public offering shares (99.85% when one plaintiff bought 3,000 shares and 91% when another bought his shares), there was no way of determining to what extent they had bought the public offering shares versus the

---

[27] See, e.g., Guenther v. Armstrong Rubber Co., 406 F.2d 1315 (3rd Cir. 1969) (holding that although 75–80% of tires marketed by Sears were made by defendant manufacturer, plaintiff would have lost on a directed verdict even if he had been injured by a tire bought at Sears).

[28] Howard v. Wal-Mart Stores, Inc., 160 F. 3d 358, 359–60 (7th Ci. 1998) (Posner, C.J.).

[29] 402 F. 3d 489 (5th Cir. 2005).

insider shares. Assuming random selection from the pool, plaintiffs' expert computed the probability that at least one share would have been from the public offering segment and the court did not dispute that this probability was virtually 1. The court acknowledged that the burden of proof of standing was by a preponderance of the evidence and that all evidence is probabilistic, but nevertheless ruled that probabilities were not enough, citing *Sargent, Smith,* and Judge Posner's dictum. In particular, the court concluded that allowing the probabilistic argument would contravene the statute by extending it to all purchasers of shares from the pool, since all would have the same argument. It also rejected the "fungible mass" argument by which each purchaser would be deemed to have purchased his proportionate share of public offering shares; allowing that argument, the court held, would undermine the requirement that plaintiff trace his shares to the public offering. When the act was passed, in 1933, Wall Street practices were different and shares could in many cases be traced. Since that was no longer true, the remedy, the court concluded, was to amend the statute.

The court did not note that in *Krim* no other evidence would have been possible to identify the source of the shares, so the case might have been held to fall within Judge Posner's caveat holding statistics sufficient in such circumstances. Also, the hypothetical Judge Posner discussed was a bare statistical preponderance of 51%, far from the overwhelming probability in *Kim*. The *Krim* court complained that if it allowed plaintiffs' standing, all purchasers from the pool could sue, a result it found inconsistent with the Securities Act, which required tracing. But under the court's ruling, no purchaser could sue, a barrier to relief surely in conflict with the remedial purposes of the Securities Act. As for economic advantage (Judge Posner's argument for holding statistics insufficient in the bus case), this assumes there are many such cases. If accidents involving unidentified buses become numerous enough to be a significant economic burden, the better solution would seem to be an application of enterprise liability, the ultimate statistical justice, in which each company bears its proportionate share of liability in each case.

Those who find "bare" statistical evidence intrinsically insufficient must explain why it somehow should become magically sufficient if there is even a smidgeon of case-specific evidence to support it.

Fortunately the utterances in the early cases are only dicta that have not been uniformly solidified into bad rules of evidence or law. In most situations, the actual rules are much sounder from a Bayesian point of view. The learned Justice Lummus may disparage statistical evidence by writing that the fact that only a minority of men die of cancer would not warrant a finding that a particular man did not die of cancer, and yet in an earlier case appraise the force of statistical proof by soundly holding that "the fact that a great majority of men are sane, and the probability that any particular man is sane, may be deemed by the jury to outweigh, in evidential value, testimony that he is insane . . . [I]t is not . . . the presumption of sanity that may be weighed as evidence, but rather the rational probability on which the presumption rests."[30]

---

[30] Commonwealth v. Clark, 292 Mass. 409, 415 (1935) (Lummus, J.) (citations omitted).

When the event is unusual, but not so rare as to justify a presumption against it, we may fairly require especially persuasive proof to overcome the weight of statistical or background evidence. That, arguably, is the reason for the rule requiring "clear and convincing" evidence in civil cases in which the claimed event is unusual, as in cases involving the impeachment of an instrument that is regular on its face. By this standard we stipulate the need for strong evidence to overcome the negative background. When the statistics are very strong and negative to the proposition asserted, particular proof may be precluded altogether. In the Agent Orange, Bendectin, and silicone-breast-implant litigations some courts refused to permit experts to testify that defendants' products caused plaintiffs' harms, since the overwhelming weight of epidemiological evidence showed no causal relation.[31] It is entirely consistent with Bayes's theorem to conclude that weak case-specific evidence is insufficient when the background evidence creates very strong prior probabilities that negate it.

It is one thing to allow Bayes's theorem to help us understand the force of statistical evidence, it is another to make explicit use of it in the courtroom. The latter possibility has provoked considerable academic and some judicial debate. In paternity cases, where there is no jury, some courts have permitted blood typing experts to testify to the posterior probability of paternity by incorporating a 50% prior probability of paternity based on the non-blood-type evidence in the case. But in one case a judge rejected the testimony and recomputed the posterior probability because he disagreed with the expert's assumed prior.[32]

The acceptability of an express use of Bayes's theorem was explored in a New Jersey criminal case, in which a black prison guard was prosecuted for having sexual intercourse with an inmate, which was a crime under state law. She became pregnant and the child's blood type matched that of the guard's. At the trial an expert testified that 1.9% of black males had that blood type and so the exclusion rate was 98–99%. Using a 50% prior she testified that there was a 96.5% probability that the accused was the father. On appeal, the intermediate appellate court reversed the conviction, quoting from an opinion by the Wisconsin Supreme Court: "It is antithetical to our system of criminal justice to allow the state, through the use of statistical evidence which assumes that the defendant committed the crime, to prove that the defendant committed the crime."[33] This is clearly incorrect because the prior in Bayes's theorem does not assume that the accused committed the crime, but only that there was a probability that he did so. The court was right, however, in refusing to let the expert testify based on his prior because he had no expertise in picking a prior and because testimony based on his prior would not be relevant for jurors who had different

---

[31] In re Agent Orange Product Liability Lit., 611 F. Supp. 1223, 1231–1234 (E.D.N.Y 1985), aff'd, 818 F.2d 187 (2d Cir. 1987) (Agent Orange used in Vietnam and various ailments); Daubert v. Merrell Dow Pharmaceuticals, Inc., 43 F.3d 1311 (9th Cir. 1995) (Bendectin and birth defects); Hall v. Baxter Healthcare Corp., 947 F. Supp. 1387 (D. Ore. 1996) (silicone breast implants and connective tissue disease).

[32] Hawkins v. Jones, Doc. Nos. P-2287/86 K and P-3480/86 K (N.Y. Family Ct. Jan. 9, 1987).

[33] "Paternity Test at Issue in New Jersey Sex-Assault Case," The New York Times November 28, 1999 at Bl.

priors. On final appeal, the Supreme Court of New Jersey affirmed the intermediate court's reversal of the conviction, suggesting that the expert should have given the jurors posterior probabilities for a range of priors so they could match their own quantified prior views with the statistical evidence.[34] If the expert cannot use his own prior, is the New Jersey Supreme Court right that she can give jurors a formula and tell them to insert their prior, or give them illustrative results for a range of priors? In addition to New Jersey, one other state supreme court has suggested that an expert may give illustrative results for a range of priors.[35] At least one court has held otherwise.[36] The U.K. judges have explored the issue more fully and ruled out an explicit use of the theorem at trial, noting that "To introduce Bayes Theorem, or any similar method, into a criminal trial plunges the Jury into inappropriate and unnecessary realms of theory and complexity, deflecting them from their proper task."[37]

Perhaps the strongest case for an explicit use by the prosecution arises if the defense argues that the trace evidence does no more than place the defendant in a group consisting of those in the source population with the trace in question. The prosecution might then be justified in using Bayes's theorem to show what the probabilities of guilt would be if the jurors believed at least some of the other evidence. Conversely, if the prosecutor argues that the probability of guilt is the complement of the frequency of such traces in the population (thus assuming a 50% prior probability of guilt), the defense may then be justified, using Bayes's theorem, to demonstrate what the probabilities would be if some or all of the other evidence were disbelieved.

Another, perhaps better way of proceeding, is to allow the expert to describe the effect of the likelihood ratio associated with the test without making any assumption about the prior probabilities. Thus the expert could testify that we are a thousand times more likely to see a matching print if defendant left it than if someone else did. Or, alternatively, the odds that defendant left the print are increased a 1,000-fold by finding a match. In both formulations the expert makes no assumption about prior probabilities, but only testifies to the *change* in probabilities or odds associated with the test. One expert panel (of which I was a member) has endorsed this approach.[38]

Whichever way the issue of explicit use of Bayes's theorem is resolved, what is more important is the larger teaching that is itself often ignored: A matching trace does not have to be unique or nearly unique to the defendant to make a powerful case when combined with other evidence.

---

[34] State v. Spann, 130 N.J. 484 (1993).

[35] Plemel v. Walter, 303 Ore. 262 (1987).

[36] Connecticut v. Skipper, 228 Conn. 610 (1994).

[37] Regina v. Dennis John Adams, EWCA Crim 222, transcript at *13 (April 26, 1996), quoted with approval in Regina v. Alan Doheny & Gary Adams, [1996] EWCA Crim 728, transcript at *4 (July 31, 1966) (dictum).

[38] Committee on Scientific Assessment of Bullet Lead Elemental Composition Comparison, *Forensic Analysis: Weighing Bullet Lead Evidence*, 96, 97, 112 (The National Academies Press, 2004).

Chapter 2
Descriptive Tools

Description by summary is a basic function of statistical work. The essential tools are *measures of central location* – principally the mean, median, and mode – that locate in various ways the center of the data; *measures of variability* – most notably the variance and standard deviation – that express how widely data observations vary around their central value; and *measures of correlation* – particularly Pearson's product–moment correlation coefficient – that describe the extent to which pairs of observations are linearly related. These are the building blocks on which the edifice of statistical reasoning is based. Let us look at their definitions in some detail.

Measures of Central Location

Mean

The mean is the most common measure of location for data. In its simplest form it is the familiar arithmetic average, which is the sum of the observations divided by their number. The mean is a "central" value in the following senses:

- The mean is the unique number for which the algebraic sum of the differences of the data from that number is zero (the sum of deviations above the mean equals the sum of deviations below the mean).
- Consequently, if each value in the data is replaced by the mean, the sum is unchanged.
- The sum of the *squared* deviations from the mean is smaller than for any other value from which squared deviations might be measured. (This is an important characteristic used in statistical models to be discussed.)

The mean is a central value as described above, but it is not necessarily a "representative" or "typical" value. No one has the average 2.4 children. Nor is the mean necessarily the most useful number to summarize data when a few large (or small) observations can be important. One can drown in a river that has an average depth of 6 inches or have been felled by the heat in July 1998, when the Earth's average

M.O. Finkelstein (ed.), *Basic Concepts of Probability and Statistics in the Law*,
DOI 10.1007/b105519_2, © Springer Science+Business Media, LLC 2009

temperature was 61.7°F. The fact that a woman's life expectancy exceeds a man's is not inconsistent with the fact that many women will not outlive the male average life. That fact led the U.S. Supreme Court to hold that charging women more for pensions was discriminatory because "[e]ven a true generalization about the class is an insufficient reason for disqualifying an individual to whom the generalization does not apply."[1]

The sample mean is a useful estimator of the population mean because as an estimator it is both *unbiased* and *consistent*. A sample estimator is said to be unbiased if its average over all possible random samples equals the population parameter (here, the mean), no matter what that value might be. In short, the estimator is not systematically off the mark. An estimator is said to be consistent if, as the sample size increases, the probability that the sample estimator will differ from the population parameter by any given amount approaches zero. (This is a statement of the law of large numbers, discussed in Chapter 1.) The sample mean has both characteristics, but other estimators may have only one or neither. Of the two properties, consistency is the more important because, when an estimator is consistent, any bias becomes negligible as sample size increases.

Median

The *median* of data is any value such that at least half the values lie at or above it and at least half lie at or below it. When there are an even number of observations, it is usual to take the median as midway between the two middle observations when the observations are ranked from smallest to largest. When there are an odd number of observations, the median is the middle observation. The median is a commonly used statistic when the distribution of observations is highly skewed. Income distributions are a prime example. To take an extreme case, if four people have annual incomes of $10,000 each and one person has an income of $1,000,000 the mean income would be $208,000, but the median income would be $10,000. Neither statistic is a good summary of this highly skewed data set, but the median seems to be closer to the center of the data than the mean, which is overly influenced by one highly discrepant observation. For this reason, medians are frequently used with skewed distributions, such as income or family size. In such cases the mean, pulled by large observations, will be above the median. See Fig. 2.1 for an illustration.

One property of the median is that the sum of the absolute deviations from the median is smaller than for any other central value from which deviations might be measured. When dealing with sample data, the mean is preferred by statisticians because generally it will have a smaller sampling error than the median: i.e., in repeated random sampling from a population, the sample mean will vary less from sample to sample than the sample median. This characteristic has not, however, been a factor in the choice between the mean and the median in the law.

---

[1] City of Los Angeles Dep't of Water and Power v. Manhart, 435 U.S. 702, 708 (1978).

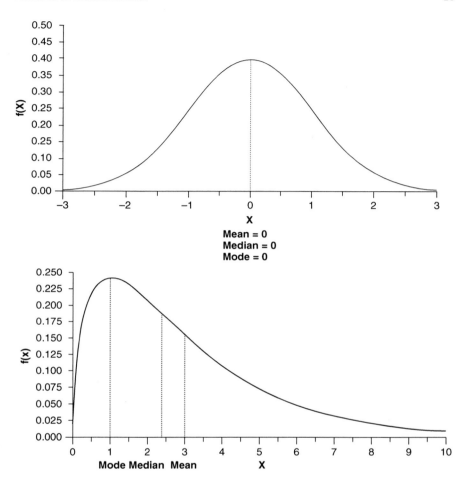

Fig. 2.1 Symmetrical and asymmetrical distributions

Here are two cases in which the courts favored the median because they thought the mean was too highly influenced by a few large values.

To protect railroads from discriminatory taxation by state and local governments, an act of Congress prohibited state and local authorities from assessing railroads for property taxes at higher rates than those for other types of commercial and industrial properties. (The rate in question is the ratio of assessed value to market value; the average ratio is determined from recent sales of properties.) Was the right measure of comparison the mean ratio of assessment to market value or the median ratio? One court favored the mean, apparently on the ground that the legislative history referred to the "average" ratio, and the term "average" could only refer to the mean.[2] But another court, focusing more persuasively on the purpose of the statute, favored the median because, it concluded, the mean would be overly influenced by large properties held by public service companies that were centrally assessed at higher rates in much the same way that railroads were assessed.[3] The court argued that, under the act, Congress intended that railroads should not be compared primarily with utilities, which could be subject to similar local prejudices, but more broadly with other industrial and commercial properties.

In another case, the estate of a patient who died from a virulent form of brain cancer sued the decedent's doctors for wrongful death because, the estate claimed, the doctors were guilty of malpractice in trying a new experimental treatment that had shortened his life. To support the claim of shortened life, plaintiff cited a study of 70 patients with the same cancer, which reported that the mean life expectancy was 17 months after first diagnosis, while the decedent had lived only 7 months. But the court pointed out that the median life in the study was 8.3 months, a figure the study's authors said was more representative because the average life was distorted by a few unusually long-lived cases.[4] Assuming, as the parties did, that the study cases reflected the deceased's life expectancy in the absence of the experimental treatment, he had about a 50% chance to at least equal the median life, but less than a 50% chance to reach the mean life. So it could not be said, by a preponderance of the evidence, that he would have lived longer than the median without the experimental treatment. (The court quite reasonably concluded that the difference between the deceased's life of 7 months and the median life of 8.3 months was too small to matter.)

On the other hand, there are cases in which one wants to take account of all the data even if the distribution is skewed by some large values; in such cases the mean is preferred to the median as a measure of central location. Here is an odd example in which the usual direction of inference in statistics is reversed: We know about the population and seek to make inferences about a sample.

An employee of Cisco was convicted of embezzling some 150,000 shares of Terayon stock owned by Cisco. Cisco eventually recovered the shares, but was injured by being deprived of the opportunity to sell them over a 19-month period. Cisco argued that its measure of loss was the mean daily closing price of Terayon shares over the period times the number of shares; this came to about $12.5 million. The employee argued, among other things, that the median price should be used, in which case the loss would be substantially less. The district court found that there was no way to determine when the embezzled shares would have been liquidated by Cisco, except that all the shares would have been sold by the end of the period.

---

[2] Southern Pacific v. California State Board of Equalization, unpublished decision, discussed in D. Freedman, "The Mean Versus the Median: A Case Study in 4-R Act Litigation," 3 Journal of Business & Economic Statistics 1 (1985).

[3] Clinchfield R.R. Company v. Lynch, 700 F.2d 126 (4th Cir. 1983).

[4] Heinrich v. Sweet, 308 F.3d 48, 62 (1st Cir. 2002), cert. denied, 539 U.S. 914 (2003).

If Cisco had sold all the shares on the same day, the argument for the median price would have been a good one; the case would have been like the brain tumor death. But this would have been unlikely, given the market disruption that selling such a large block of shares all at once would have caused. If, as was likely, the shares would have been sold over a significant number of days, the law of large numbers would have insured that the mean closing price for those days (the sample mean) would have closely approximated the mean price over the whole period (the population mean), so that the latter would have been a good estimator of the former. Finally, substitution of the sample mean price for daily prices in the sample would be justified because the sum of the closing prices on those days would not be changed by substituting the population mean price for each day.

The district court, without discussion, went with the mean and the higher restitution amount; its decision was affirmed on appeal.[5]

The median may be easier to determine than the mean because one only needs half the data in an ordered set to calculate the median, but all of the data to calculate the mean. This difference becomes important when mean lifetimes are sought but not all members of the population have died.

Mode

The mode is the most frequent value in a set of data. It is used in *maximum likelihood estimation*, in which the statistician adopts as an estimate for some unknown parameter the value that would make the observed evidence most likely to appear. Maximum likelihood estimates are an important tool in more advanced statistics.

Variants of the Mean

It is sometimes appropriate to use a *weighted mean*, as for example when an overall figure is to be derived from figures for strata of different sizes. In that case the mean for each stratum is weighted proportionally to the size of the stratum. In the railroad example, the average assessment/sales ratio for other properties was computed by weighting the ratios for the sample properties by their sizes. This was noncontroversial.

But whether to use weighted or unweighted means can be a subject of controversy. In another case, relying on a sleep study, a drug manufacturer claimed that taking its antihistamine reduced time to fall asleep ("sleep latency") by 46%. The company reached this figure by adding up the sleep latencies for the study subjects after they took a pill and dividing that by the sum of the subjects' sleep latencies without a pill (the baseline). This was equivalent to a weighted average latency, with the weights proportional to the baseline latencies. The Federal Trade Commission

---

[5] United States v. Gordon, 393 F.3d 1044 (9th Cir. 2004), *cert. denied*, 546 U.S. 957 (2005).

challenged the company's calculation. It got a smaller figure by computing the percentage reduction in sleep latency for each subject and averaging those percentages. This was an unweighted average. The reason for the difference in result was that subjects who had the greatest baseline sleep latencies experienced the greatest percentage reductions when they took a pill; using the weighted average gave greater effect to these long-latency subjects compared with a simple unweighted average. Which figure is "right"? The weighted figure would seem to be preferable because it reflects the percentage reduction for the average baseline latency. But the larger point is that one cannot fairly talk about a percentage reduction in sleep latency without specifying the baseline latency.

Another case involved the interpretation of a federal statute that required state medicaid plans (which cover payments for poor persons) to identify hospitals that had disproportionate numbers of such patients and to provide for incremental payments to such hospitals. The federal statute defined a disproportionate-share hospital as one in which "the hospital's medicaid inpatient utilization rate [the percentage of the hospital's inpatient days that were attributable to medicaid patients]... is at least one standard deviation above *the mean* medicaid inpatient utilization rate for hospitals receiving medicaid payments."[6] In computing the mean for these purposes California used a mean weighted by total hospital inpatient days; this had the effect of raising the eligibility threshold because large urban hospitals had higher proportions of medicaid patients than small and rural hospitals. A hospital that was not eligible under this definition sued, claiming that the federal statute, by specifying a mean, precluded using a weighted mean; it argued that the state agency should treat all hospitals alike and as a single data point in computing the arithmetic mean number of medicaid inpatient days. The court rejected this argument. It held that the statutory term "mean" was ambiguous and could include either a weighted or an unweighted mean. This seems correct. According to the court, there was evidence that Congress intended to permit the former because it had subsequently amended the statute in another place, to include the phrase "arithmetic mean," but had not made a change in this provision. Since the statute was ambiguous and the responsible federal agency had acted reasonably in approving the use of the weighted mean, the court would defer to the agency.[7]

When the mean would be skewed by highly deviant observations, variants may be used that may improve its representativeness as a measure of the center of the data. One of these is the *trimmed mean*, which is the mean after discarding the top and bottom *P* % [e.g., 5%] of the data. Another is the *geometric mean*. The geometric mean of *n* positive numbers is the *n*th root of their product. Perhaps a more intuitive definition uses logarithms: the arithmetic average of the logarithms of the data is the logarithm of the geometric mean. Taking the anti-log recovers the geometric mean.

---

[6] 42 U.S.C. § 1396r-4(b)(1)(A)(2006)(emphasis supplied).

[7] Garfield Medical Center v. Kimberly Belshe, 68 Cal. App. 4th 798 (Cal. App. 1998). I would think that the term "arithmetic mean" would not unambiguously point to an unweighted mean, as the court assumed in its argument from legislative history.

The log scale deemphasizes differences at the high end of a scale compared with equal differences at the low end. The difference between 1 and 101 is 100 and so is the difference between 1,000 and 1,100. But in the log scale the first is about 2, while the second is about 0.04. By giving less weight to differences at the high end, the geometric mean is always less than the arithmetic mean when all numbers are positive. For example, the arithmetic mean of 1 and 100 is 50.5; the geometric mean is 10.

The geometric mean is encountered in regulations incorporating some average when the data are bounded by zero on one side and unbounded on the other, so that a skewed distribution is likely to result, but the regulators do not want an average that is unduly influenced by a few large values. Length of hospital stays in medicare reimbursement rules and counts of fecal coliform bacteria in water quality standards are examples of the use of geometric mean for this reason. When, in 1998, the consumer price index, as computed by the Bureau of Labor Statistics, was criticized as overstating increases in the cost of living, the Bureau responded by adopting the geometric mean to compute average price levels across related products. This had the effect of damping down the rise in the index when one product increased in price but others in the group stayed at or near the same levels.

A limitation of the geometric mean is that while the arithmetic mean in a sample can be used to extrapolate to a population total by multiplying by the inverse of the sampling fraction, the geometric mean cannot be so used. In one case the issue was the number of fish killed per year by a hydroelectric dam.[8] This number was projected from a daily average, based on random days, multiplied by 365. The defendant company argued that because the daily number of fish killed varied widely, and on some days was very high, the geometric mean number of fish killed per day should have been used instead of the arithmetic mean. But this could not be done because there is no way to go from the geometric daily average to a yearly total. (As one example, note that if on any day included in the sample, zero fish were killed, the geometric daily average could not be computed because it is defined only for positive values.)

Another variant of the mean is the *harmonic mean*. Its definition sounds arcane: it is the reciprocal of the average reciprocals of the data. It is used when small values must be weighted more heavily than large values in taking an average. Defined for positive values, it is always less than the geometric mean. For example, the Environmental Protection Agency prescribes that if a vehicle is given more than one fuel economy test, the harmonic mean of the test results shall be taken as representing the vehicle.[9] To take an extreme case, suppose that in one test the vehicle had a mileage figure of 20 mpg and in another had 40 mpg. The average of the reciprocals would be $\frac{1}{2}\left(\frac{1}{20} + \frac{1}{40}\right) = \frac{3}{80}$; the harmonic mean would be the reciprocal of that, or $80/3 = 26.6$ mpg. The geometric mean is 28.3 mpg and the arithmetic mean is

---

[8] Kelley *ex rel.* Michigan Dep't of Natural Resources *ex rel.* v. FERC, 96 F.3d 1482 (D.C. Cir. 1996).

[9] EPA Review of fuel economy data, 40 C.F.R. § 600.008-01 (2005).

30 mpg. Which measure is "right"? Assume that the vehicle was driven 100 miles in each test. Then the total gas consumed would be $\frac{100}{20} + \frac{100}{40} = 7.5$ gal; this is $200/7.5 = 26.6$ mpg. Note that in computing an overall figure, reciprocals of the mpg are used, the low value is given more weight, and the harmonic mean gives the correct answer; both the geometric and arithmetic means give answers that are too high.

Although the harmonic mean is always less than the geometric mean, the two tend to be close, so the difference between them, in many cases, may not be all that important. But in one major constitutional case, it was at the center of the lawsuit. The U.S. Constitution requires the apportionment of Representatives among the States "according to their respective Numbers." Ideally, under the one-person, one-vote standard, the number of persons per representative would be the same in each state. Exact equality, however, is not possible given that the number of representatives is now fixed by statute at 435, each state must have at least one representative, and fractional representatives are not allowed. After allocating one representative to each state, the general method is to allocate the 51st representative to the state with the largest population, the 52nd representative to the state with the largest population "per representative" (in a sense to be described), and so on until all 435 representatives have been assigned. But in calculating the population "per representative" (called the state's priority number) does one use as the divisor the number of representatives the state already has? the number it will have if it receives the assignment? or some average of the two? Various methods were used until Congress, in 1941, acting on a recommendation of a committee of mathematicians, adopted the "method of equal proportions," also known as the Hill method. In this method, when considering the allocation of the nth seat to a state, the divisor is the geometric mean of the state's nth and $(n-1)$ seats. This method was justified because for any pair of states it results in an allocation with a smaller *relative* difference than if a seat were shifted from one state to the other. (The relative difference for a pair of states is the ratio of the number of persons per seat in whichever state has the larger number to the number of persons per seat in the other state.)

In the 1990 apportionment, Montana went from two seats to one, losing a representative to the State of Washington, which went from eight to nine seats. The reason was that in allocating the last seat, Montana's priority number was computed using the geometric mean of 1 and 2, or $\sqrt{(1)(2)} = 1.41$, as required by statute. Montana sued, claiming that the harmonic mean (also known as the Dean method) should have been used instead, since that would better carry out the constitutional equality standards. If the harmonic mean had been used, Montana's priority number would have been computed using the divisor $\frac{1}{2}\left(\frac{1}{1} + \frac{1}{2}\right)^{-1} = 1.33$. The slightly smaller divisor gave it a higher priority number than Washington would have received using the harmonic mean in going from eight to nine seats. Since Montana would have the higher priority number it would have received the last seat. And, in general, use of the harmonic mean would have favored small states because the harmonic mean is always less than the geometric mean.

Montana justified the harmonic mean by pointing out that the allocation it generated would minimize the *absolute* deviation from the ideal district in any pair of states. The absolute deviation is the number of persons per representative that the state's average district departs from the ideal; the sum of these for the two states is the absolute deviation for the pair. In Montana's case, using the geometric mean produces an absolute deviation of 260,000 from the ideal, while using the harmonic mean reduces the deviation to 209,165. But by using the harmonic mean the *relative* difference is increased to 1.52, which means that Washington is 52% worse off than Montana. Which measure of inequality – the relative or absolute difference – was constitutionally compelled? The answer is neither. The Supreme Court unanimously rejected Montana's challenge, holding that "i[i]n none of these alternative measures of inequality [the absolute difference, the relative difference, and others] do we find a substantive principle of commanding constitutional significance."[10] In short, Congress could have used either the harmonic mean, the geometric mean, or others, and the statute prescribing the geometric mean was not unconstitutional for making that choice among possible means.

Sometimes even the mean and the median are not enough to tell the story. For example, the mean U.S. gross domestic product per capita is far greater than Sweden's, and the median American family has a standard of living that is roughly comparable to the median Swedish family. But that doesn't mean that Americans of all classes are better off than, or even as well off as, their Swedish counterparts. As one moves down in income class, Swedish living standards pull far ahead of U.S. standards: The bottom 10% of Swedes have incomes that are 60% higher than those of their U.S. counterparts. The United States has a higher GDP per capita than Sweden because America's rich are richer than Sweden's rich.[11]

Measures of Dispersion

Summary measures of dispersion (variability or scatter) are important descriptive devices because they indicate the degree of deviation in the data from the central values. Like measures of central location, dispersion measures come in various forms that emphasize differing aspects of the underlying data.

Variance and Standard Deviation

By far the most important measures of dispersion are the *variance* and the *standard deviation*. The variance is the mean squared deviation of the data from their mean value and the standard deviation is the positive square root of the variance. For data they are usually denoted by σ^2 for the variance and σ for the standard deviation. These quantities measure dispersion because the further the data points lie from the

---

[10] U.S. Department of Commerce v. Montana, 503 U.S. 442, 463 (1992).

[11] P. Krugman, "For Richer: How the Permissive Capitalism of the Boom Destroyed American Equality," The New York Times Magazine 62, 76 (October 20, 2002).

mean value the larger the variance and standard deviation. The following properties
follow immediately from the definitions:

- Variance and standard deviation are never negative, and are zero only when the
  data have no variation.
- Adding a constant to the data does not change the variance or standard deviation.
- Multiplying the data by a constant, c, multiplies the variance by c^2 and multiplies
  the standard deviation by the absolute value of c.
- The standard deviation is in the same units as the data. From this follows a useful
  property: In measuring how far a data point is from the mean, the units don't matter; a man who is one standard deviation above the mean height is still one standard deviation above it, whether heights are measured in centimeters or inches.

When the variability of data is a significant issue, the standard deviation may be
added to the mean and median to describe the distribution. For example, in one case,
Baxter International reported projections of future earnings that proved to be much
too rosy; when the actual results were reported the price of the company's stock
declined substantially and Baxter was sued for misleading projections. The court
held that Baxter must have made a variety of internal estimates and that to give the
markets sufficient information to generate an accurate stock price it should have
disclosed the mean, median, and standard deviation of those estimates. The court
observed that "Knowledge that the mean is above the median, or that the standard
deviation is substantial would be particularly helpful to those professional investors
whose trades determine the market price."[12]

In the famous Exxon Valdez case the Supreme Court considered the validity of a
$2.5 billion dollar punitive damage award against Exxon for spilling millions of gal-
lons of crude oil into Prince William Sound in Alaska.[13] In reducing the award, the
Court noted the unpredictability of punitive damages and cited as evidence of that a
comprehensive study of such awards in state civil trials. The median ratio of punitive
to compensatory awards was 0.62:1, but the mean ratio was 2.90:1 and the standard
deviation was 13.81. From this the Court concluded that "the spread is great, and the
outlier cases subject defendants to punitive damages that dwarf the corresponding
compensatories."[14] This conclusion was unwarranted in Exxon's case because the
analysis did not take into consideration the size of the compensatory awards, and the
standard deviation of punitives for large compensatory awards, like Exxon's, was far
smaller than the standard deviation over all awards, the figure used by the Court.[15]

---

[12] Asher v. Baxter Int'l, Inc., 377 F.3d 727, 733 (7th Cir. 2004). One must say, however, that the
court's observation was no more than a speculation; it assumed there were other internal estimates
and that the median of those estimates would have been sufficiently below the mean (which presumably was the projection that was reported) to suggest that the mean was unrepresentative.

[13] Exxon Shipping Co. v. Baker, 128 S. Ct. 2605 (2008).

[14] *Id.* at 2625.

[15] I am indebted to Professor Theodore Eisenberg for calling this to my attention.

The standard deviation is particularly useful as a measure of variability because it can be used to make important statements about the distribution of data. A prime example is the practice of giving estimates plus or minus two standard deviations to allow for uncertainties in estimation or measurement. This derives from the frequently made assumption that the data have a normal distribution (the well known bell-shaped curve, which is discussed in Chapter 5), and in that case only about 5% of the data lie beyond 1.96 standard deviations from the mean. Thus, for example, when the FBI sought to connect a bullet found at a crime scene with a bullet found in the home of a suspect, it followed the practice of measuring trace elements in both bullets and declaring a match when the measurements were within four standard deviations of each other (an allowance of two standard deviations for each measurement).[16]

When the data cannot be assumed to be normal, there is an important theorem (not usually encountered in practice) due to the Russian mathematician Pafnuty Lvovich Chebyshev (1824–1894) that gives an intuitively appealing interpretation of the standard deviation. The theorem, Chebyshev's inequality, is as follows: For any set of data, the probability that a randomly selected data point would lie more than k standard deviations from the mean is less than $1/k^2$. This means, for example, that fewer than $1/4$ of the points in any data set lie more than two standard deviations from the mean and fewer than $1/9$ lie more than three standard deviations from the mean, etc. The importance of the theorem derives from its generality: It is true of *any* data set. Notice that this is a much weaker statement than can be made when the data are assumed to be normal.

Standard deviations are also commonly used to identify data points that are far from the mean. For example, in determining whether a child is entitled to Social Security Disability Benefits, the Commissioner must find that the child has "marked" limitations in two domains of functioning or "extreme" limitation in one domain. (Domains are "broad areas of functioning intended to capture all of what a child can or cannot do.") The regulations provide that a child's limitation is "marked" if he or she scores at least two standard deviations below the mean on standardized testing, and is "extreme" if the score is three or more standard deviations below the mean. Assuming a normal distribution of test scores around the mean (which is not unreasonable), the child with a "marked" limitation would be no higher than the bottom 2.5% in functioning and with an "extreme" limitation would be no higher than the bottom 0.1%.[17]

The fact that observations more than two standard deviations from the mean are deemed to be marked departures from the average does not mean that values within two standard deviations are comparable to the average. There is plenty of room for

---

[16] The FBI's method of declaring a match is not technically correct, although it is close. See M.O. Finkelstein & B. Levin, "Compositional Analysis of Bullet Lead as Forenic Evidence," 13 Journal of Law and Policy 119, 123 (2005). Under criticism for overstating the probative significance of its results, the FBI discontinued making such analyses in 2006.

[17] 20 C.F.R. § 416.926a(e)(2)(iii); Briggs *ex rel.* Briggs v. Massanari, 248 F.3d 1235 (10th Cir. 2001).

variation within two standard deviations, and whether observations within that range are comparable obviously depends on the context. In 1992, the Federal Communications Commission issued a rule defining rates for telecommunications services in rural areas as "reasonably comparable" to rates in urban areas when they were within two standard deviations of the national average urban rate. When this definition was challenged in court, the FCC defended by pointing out that the highest urban rate would be near the two-standard-deviation benchmark and so a rural rate that was at the benchmark would be close to the highest urban rate, i.e., comparable to it. The court acknowledged the logic of this approach, but rejected it. Among other things, the court noted that the comparability benchmark rate would be 138% of the national average urban rate and more than twice the lowest urban rate. The court concluded that "we fail to see how they [the benchmark and national average urban rates] could be deemed reasonably comparable...."[18]

While the variance and standard deviation are used almost reflexively by statisticians, other measures of variability may be more appropriate in certain legal contexts. As in the case of the mean, problems may arise when there are significant outliers (data points far from the main body). Because the variance uses squared differences, it gives greater weight to large deviations and this may seriously distort the measure of dispersion. Moreover, quite apart from the distorting effect of outliers, the problem at hand may be one for which squared deviations are simply inappropriate as a description.

The simplest class of alternative measures involves the difference between upper and lower percentiles. The *range* of a data set is the difference between the largest and smallest number in the set. Obviously, the range is highly sensitive to outliers. Less sensitive is the *interquartile range*, which is the difference between the 25th and 75th percentiles and thus gives an interval containing the central 50% of the data. Another measure of dispersion, the *mean absolute deviation*, is the average absolute magnitude of departures from the mean value. Since deviations are not squared, the mean absolute deviation will usually be smaller than the standard deviation.

Measurement of dispersion was a prominent feature of the Supreme Court's early reapportionment cases. Under the one-person, one-vote doctrine, variances in the population of congressional or state electoral districts are allowed to only a limited extent. For congressional districts, the allowance is "only the limited population variances which are unavoidable despite a good-faith effort to achieve absolute equality, or for which justification is shown."[19] For state reapportionment, there is a less strict standard under which small divergences from a strict population equality are permitted, but only if "based on legitimate considerations incident to the effectuation of a rational state policy."[20]

Under these standards, what summary measure of variation should judges use in cases challenging reapportionment? Statisticians might instinctively choose the standard deviation, but there is no particular reason to prefer a measure based on

---

[18] Qwest Communs. Int'l, Inc. v. FCC, 398 F.3d 1222, 1237 (10th Cir. 2005).

[19] Kirkpatrick v. Preisler, 394 U.S. 526, 531 (1969).

[20] Reynolds v. Sims, 377 U.S. 533, 579 (1964).

squared deviations, which in this context has no intuitively appealing interpretation. The Supreme Court has looked at the range (the difference between the largest and smallest districts in their departure from the ideal district), the mean absolute deviation (referred to as the average deviation), and variations on the interquartile range (e.g., number of districts within X percentage points of the ideal) to measure disparity. This seems unobjectionable.

Much more questionable from a statistical point of view was the Supreme Court's concern with electoral variability in light of the one-person, one-vote doctrine that was the ostensible basis for its famous decision in *Bush v. Gore*.[21] As everyone knows, in the 2000 presidential election the Court stopped the manual recount of ballots ordered by the Florida Supreme Court. The Court's *per curiam* opinion (7–2) observed that in determining whether the intent of the voter could be discerned in the manual recount (the Florida Supreme Court's standard), some voters might have had their ballots judged by stricter standards than others. As an example, the Court noted that in Broward County, which had already completed its manual recount, 26% of the uncounted votes had been recovered, i.e., the voters' intent could be discerned, while in Palm Beach County, which had also manually recounted its votes, only 8% of the uncounted votes had been recovered. Seven justices agreed that the possibility that different counties (or even different precincts) could apply different standards in determining the intent of the voter in manually reviewing undervote ballots raised an equal-protection issue.

The argument can be spelled out as follows: A voter is denied equal protection by variation in county standards if her vote was not counted but would have been had she voted in a different county with a more lenient standard and a higher recovery rate. The number of voters so hurt can be viewed as a measure of the equal-protection problem. This number is linked to the rates of undervoting and the variation in such rates among the counties in the sense that if the rates and variation in rates of undervoting among counties are decreased there are likely to be fewer people hurt in that way.

What the justices seem to have ignored was that prior to the manual recount there existed substantial variation in undervoting among the counties. Therefore, a manual recount, by reducing the number of undervotes, would likely have reduced that variation, not increased it, unless the different standards for recovery of votes applied by the counties were so large as to overcome that effect. Otherwise, in terms of reducing the rate and variation in undervoting, a recount would have served the interest in equality, not diminished it.

The numbers look like this. In the 15 counties with punch-card ballots, which were the focus of the dispute, prior to any recount, the rate of undervoting (the number of undervotes per 100 certified votes) ranged from 0.86 to 3.04 with a mean of 1.770 and a standard deviation of 0.644. Thus the standard deviation was not trivial relative to the mean. Now assume that each of the 15 counties had either a 26 or an 8% recovery rate – these being the rates cited by the Court as creating, or at least illustrating, the equal-protection problem. A computer calculation of all

---

[21] 531 U.S. 98 (2000)(per curiam).

possible distributions of the 8 and 26% recovery rates among the 15 counties (there are 32,768 of them) shows the grand mean of the mean undervote rates declining to 1.47 and the mean standard deviation of the undervoting rates declining to 0.557. In 98% of the distributions the standard deviation of the rates declines.

Allowing a recount would, therefore, probably have served the interests of equality even under a broad standard that permitted substantial variation in recovery rates. Despite an army of lawyers, this dimension of the case was not called to the justices' attention in *Bush v. Gore*.[22]

Variance of Sample Sums and the Sample Mean

The most important theoretical use of the variance and standard deviation is to measure the potential error in estimates based on random samples. If we want to estimate the average height of men in a population we could draw a random sample from the population, compute the average height in the sample, and use that as an estimate of the average height in the population. But, of course, in repeated random samples the average height in the sample would vary, so this estimator has a variability, or *sampling error*. The measure commonly used to reflect this variability is the standard deviation of the sample mean, which is called the *standard error* of the sample mean.

To derive the standard error of the sample mean, consider the standard deviation of a single draw from a population. In repeated single draws, the variance of the height of the individual drawn will equal the variance of height in the population. Denote the variance of height in a population as σ^2 and the standard deviation as σ. Now it can be shown mathematically that the variance of a sum of *independent* random variables is equal to the sum of their variances.[23] So if we select two individuals and add their heights, the variance of the sum will equal the sum of the variances of their heights. Since the variances are equal, the variance of the sum is simply two times the variance of height in the population. By extension, if we draw n individuals and add their heights, the variance of the sum will be $n\sigma^2$, and the standard error of the sum will be $\sqrt{n}\,\sigma$. If we multiply the sum by $1/n$ to get the mean, the variance of the sum, $n\sigma^2$, is multiplied by $1/n^2$, so the variance of the sample mean height is equal to the variance of height in the population divided by the sample size, or σ^2/n. The standard error of the sample mean height is simply the positive square root of that, or $\sigma/\sqrt{n}$. In words, the standard error of the sample mean height is the standard deviation of height in the population divided by the square root of the sample size. In the usual case, in making calculations, the variance of the characteristic in the population is not known and must be estimated from the sample.

---

[22] For further details on these computer studies, see M.O. Finkelstein & B. Levin, "*Bush v. Gore*: Two Neglected Lessons from a Statistical Perspective," 44 Jurimetrics Journal 181 (2004).

[23] The variance of the sum of two positively correlated variables is equal to the sum of their variances plus two times the covariance between them.

This important result is known as the square root law. It is encouraging in that the sampling error of the sample mean can be made as small as one wishes by increasing sample size, but discouraging in that accuracy increases only by the square root of the sample size, which means that much larger samples are needed to materially reduce the sampling error of the mean height estimated from the sample.

When Sir Isaac Newton was master of the Royal Mint in Great Britain, the Mint was periodically examined for the accuracy of its coinage. This was an ancient ritual dating to the thirteenth century and was known as the trial of the Pyx. Over a period of time one coin was taken from each day's production and placed in a thrice-locked box called the Pyx (from the Greek word for box; in early ecclesiastical literature, a pyx was the vessel in which the bread of the sacrament was reserved). Every few years a trial of the Pyx was declared at which the coins were weighed as a group and their aggregate weight compared with what was required by contract with the Crown. The Master of the Mint was allowed a tolerance, called a "remedy," but if that was exceeded he was fined. For many years the remedy was about 12 grains per 1b, or about 0.854 grains per sovereign. The remedy per sovereign was about equal to the standard deviation in weight (σ) of a sovereign, with the remedy for n sovereigns being, effectively, $n\sigma$.

We can now see that this remedy, aimed at protecting the Masters of the Mint from being fined for random fluctuations, may have given them too much leeway. The standard error of the aggregate weight of the sovereigns is only $\sqrt{n}\,\sigma$, assuming that variations in weight were caused by random and independent factors operating on sovereigns selected for the Pyx. Since the remedy grew linearly with n but the standard error increased only by the square root of n, the remedy, expressed as a number of standard errors, grew with the number of coins in the Pyx, one calculation giving the remedy as 189 standard errors! It is scarcely surprising that, over eight centuries, the remedy was seldom exceeded.

Correlation

To what extent are height and weight associated in a population of men? A useful measure of association is the *covariance* between them. The covariance between two variables is defined as the average value of the product of the deviation of each variable from its mean. In other words, for each man in the population, from his height subtract the mean height for all men; this is the height deviation. Make the same calculation to obtain a weight deviation. Multiply the height deviation by the weight deviation for each man and take the average of these products. This is the covariance between height and weight for men.

Since above-average values of height and weight are associated in taller men and below-average values of height and weight are associated in shorter men, the height and weight deviations will tend to be both positive (for taller men) or both negative (for shorter men). In both cases, their average product will be positive. On the other hand, if short men tended to be obese and tall men to be skinny, the average product would be negative. If height and weight were unassociated, when height was above average, about half the time weight would be above average and half the time below. In the first case, the product of the deviations would be positive and, in the second case, negative; the average of the positive and negative deviations would tend to zero. The same is true when height is below average.

The covariance does not change when a constant is added to a variable, but it does change with a change in scale, i.e., multiplying a variable by a constant multiplies the covariance by that constant. Shifting from inches to centimeters in measuring height will change the covariance with weight. To achieve a dimensionless measure that is invariant under changes in scale (as well as location) and that is standardized to be between 0 and 1 in absolute value, we divide the covariance (cov) by the product of the standard deviations (sd) of the variables to arrive at Pearson's product–moment *correlation coefficient*. When applied to describe association in a population, the correlation coefficient is denoted by the Greek letter *rho* (ρ) and when applied to a sample is denoted by r. The formula looks like this:

$$\rho = \frac{\text{cov}(X, Y)}{sd(X)sd(Y)}.$$

The correlation coefficient takes on a maximum value of plus or minus one when one variable is a linear function of the other, the case of perfect positive or negative correlation. A large positive or negative value of r signifies a strong *linear* dependence between the variables. While there are no universal rules defining strong vs. weak associations, it is often the case in social sciences that correlations coefficients of 0.50 or more are regarded as signifying strong relationships. For example, LSAT scores are regarded as sufficiently good predictors of law school grades that law school admissions are based on them. Yet the correlation between first-year law school grade point averages and LSAT scores is usually less (and sometimes much less) than 0.60.

In deciding whether correlations are big enough to "matter," an important property of r (or its population version, ρ) that helps to interpret the coefficient is that r^2 measures the proportion of total variation of one variable that is "explained" or "accounted for" by variation in the other variable and $1 - r^2$ measures the proportion that is "unexplained." By "explained" I do not mean to imply that variation in one variable is necessarily caused by variation in the other, but only that because of the association between them a part of the variation in one variable may be predicted from the other. The higher the correlation the better the prediction. This notion is made more precise in Chapter 11 on regression.

Care must be exercised when interpreting association based on the correlation coefficient for the following reasons:

1. The coefficient may show a high value because of correlation of extreme values even though there is only a weak or even non-existent linear relation in the center of the data. Compare the top panel with the middle panel of Fig. 2.2.[24] Whether such data should be viewed as correlated for legal purposes is another question.

---

[24] Conversely, a strong linear relation in a body of data may be diminished by outliers.

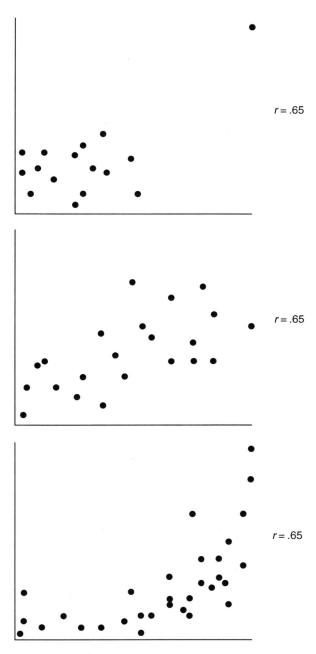

Fig. 2.2 Data patterns with r = 0.65

This situation arose in an attack on the Texas system of financing for public education as discriminatory against the relatively poor.[25] Plaintiffs sought to prove that in Texas "a direct correlation exists between the wealth of families within each district and the expenditure therein for education." In support of that proposition, plaintiffs introduced a study of about 10% of Texas school districts which compared the median family income in the district with state and local revenues per pupil spent on education. Divided into five groups by market value of taxable property per pupil, the data showed a correlation of 0.875 (treating each group as a data point) or 0.634 (weighting each group by the number of districts in the group) between revenues per pupil and median family income. However, a majority of the U.S. Supreme Court was not impressed. It noted that the data showed "only" that the wealthiest few districts in the sample had the highest family incomes and spent the most on education while the several poorest districts had the lowest family incomes and spent the least. For the remainder of the districts – almost 90% of the sample–the correlation was inverted, i.e., districts that spent next to the most on education had next to the lowest family incomes while the districts spending the least had next to the highest family incomes. From this the Court concluded that "no factual basis exists upon which to found a claim of comparative wealth discrimination."[26]

The Court's view of the data is certainly questionable. If the Court meant that the data show no correlation, it was wrong, as noted above (the correlation coefficients we have given were not calculated by the expert, nor apparently by counsel, nor by the court). The reason the overall correlation is positive despite the inversion at the center of the data is that the inverted districts are clustered near the mean, and thus convey little information about the relationship, while the districts showing a positive correlation are further removed and are thus more influential indicators. If the Court meant that the richest and poorest districts, which showed a positive correlation, were too few to brand the system, the point is hard to appraise because district sizes, in terms of numbers of pupils, were not given, and, remarkably, the professor who conducted the study did not report on how he had selected his sample of districts.

2. The correlation coefficient measures linear dependence, but may fail to reflect adequately even very strong nonlinear relations. See Fig. 2.2 (bottom panel). To take an extreme example, if X is symmetrically distributed around zero and $Y = X^2$, then it can be shown that $\rho = 0$ for X and Y, even though Y may be perfectly predicted from X. The example demonstrates that while in general if X and Y are independent then $\rho = 0$, the converse is not necessarily true, i.e., if $\rho = 0$, X, and Y are not necessarily independent. Nonlinear correlations are frequently computed using regression models.

3. Even if there is a strong (although not perfect) linear relation between two variables, X and Y, the proportion of Y's variability that is explained by X's

---

[25] San Antonio Independent School District v. Rodriguez, 411 U.S. 1 (1973).

[26] *Id.*, 411 U.S. at 27.

variability will be diminished as the observed range of X is decreased. This is called attenuation.

4. Finally, the old saw: Correlation does not necessarily imply causation. The old saw is usually illustrated with absurd examples of spurious correlations, but in real-life problems the case may be closer.

To illustrate attenuation and possibly spurious correlation consider the following Federal Trade Commission case.[27] A national trade association of egg manufacturers sponsored advertisements stating that there was no competent and reliable scientific evidence that eating eggs, even in quantity, increased the risk of heart attacks. The FTC brought a proceeding to enjoin such advertisements on the ground that they were false and deceptive. In the trial, the Commission staff introduced data prepared by the World Health Organization reporting cholesterol consumption and rates of ischemic heart disease (IHD) mortality in 40 countries. Experts who testified for the FTC subsequently collected data on egg consumption in those countries. The correlation between the two (IHD and egg consumption) is 0.426. The egg consumption and IHD data are shown in Fig. 2.3. The correlation suggests a causal association, but there are so many confounding factors that might be responsible for the correlation, such as red meat in the diet, that the evidence against eggs is at best weak. But if to eliminate confounders one limits comparisons to countries roughly similar to the United States (which is #36), at least in egg consumption, the truncation of the data causes the correlation between IHD and egg consumption to disappear, as the points within the vertical lines in Fig. 2.3 show.

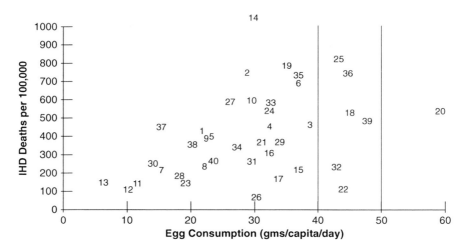

Fig. 2.3 IHD and egg consumption in 40 countries

[27] *In re* National Comm'n on Egg Nutrition, 88 F.T.C. 89 (1976), *modified*, 570 F.2d 157 (7th Cir. 1977), *cert. denied*, 439 U.S. 821 (1978).

This kind of study, which compares averages for groups (countries) to make inferences about individuals, is called *ecological*. Ecological studies are likely to have higher correlations than data for individuals because taking averages smoothes out variation within countries; with less random variation there will be higher correlations. This fact has to be kept in mind when attempting to draw conclusions for individuals from averages for groups.

Measuring the Disparity Between Two Proportions

It is remarkable how many important legal issues involve a comparison of two proportions. To make the discussion specific, consider the example of a test taken by men and women in which the pass rate for men, Y, is greater than the pass rate for women, X. Here are three candidates to describe the disparity in pass (or fail) rates.

1. The difference between pass (or fail) rates: $Y - X$

This measure has the desirable properties that (i) the difference between pass rates is the same as the difference between fail rates (except for a change in sign) and (ii) the difference in pass rates times the number of women taking the test equals the number of women adversely affected by the disparity. However, the difference may be interpreted as more or less substantial depending on the level of rates. As one court observed, "[a] 7% [percentage point] difference between 97% and 90% ought not to be treated the same as a 7% [percentage point] difference between, e.g., 14% and 7%, since the latter figures evince a much larger degree of disparity."[28]

2. The ratio of pass (or fail) rates: Y/X

This measure, called the *relative risk* or *rate ratio*, takes account of the level of rates. The relative risk is 1.08 in the 97–90% example and 2.0 in the 14–7% example. A drawback of the relative risk is that it is generally different for pass and fail rates. If one uses fail rates in the above example, the 97–90% case shows a *larger* disparity (the fail rate for women is 3.33 times the fail rate for men) than the 14–7% case (the fail rate for women is only 1.08 times the fail rate for men), thus reversing the order of disparity based on pass rates. When pass rates for both groups are high (above 0.5) there is a "ceiling effect" that constrains rate ratios to be less than 2.0, but small differences produce highly discrepant ratios between fail rates. This inconsistency is most troublesome when no clear rationale exists for preferring one rate to another.

Under the four-fifths rule of the U.S. Equal Employment Opportunity Commission, the EEOC reviews employment tests for adverse impact on any race, or other protected group, by determining whether the pass rate for that group is less than 80% of the group with the highest rate. If so, adverse impact is presumed and the

---

[28] Davis v. Dallas, 487 F. Supp. 389, 393 (N.D. Tex. 1980).

test must be validated as job related. In one case involving a proficiency test, the black pass rate was 0.582 and the white rate was 0.704; the ratio of the two (the relative risk or rate ratio) was $0.582/0.704 = 0.827$, so adverse impact would not be presumed. But the ratio of fail rates (white to black) was $0.296/0.418 = 0.708$; so if fail rates were used there would be adverse impact. If pass rates are quite high, the test becomes, in effect, one of minimum qualifications; thus one may argue that fail rates should be more significant than pass rates.

3. The odds ratio

As previously noted, the odds in favor of an event are defined as the probability that the event will occur divided by the probability that it will not. That is, if the probability of an event is p, the odds on the event are $p/(1 - p)$. So, for example, if the probability of an event is 20%, the odds on the event would be $0.20/0.80 = 1/4$, or 1 to 4. Conversely, the odds that the event would *not* occur are $0.80/0.20 = 4$, or 4 to 1.

The ratio of two odds, unsurprisingly called the odds ratio, is another statistic frequently encountered in statistical and epidemiological work. For a test taken by men and women, the odds ratio is simply the odds on passing for men divided by the odds on passing for women. The odds ratio is widely accepted as a measure of the disparity between two proportions, but it can produce some seemingly odd (no pun intended) results. For example, in the 97–90% case, the odds on passing for men are $0.97/0.03 = 32.3$; the odds on passing for women are $0.90/0.10 = 9$. The odds ratio is $32.3/9 = 3.6$ in favor of men. In contrast, the odds ratio in the 14–7% case is $(0.14/0.86)/(0.07/0.93) = 2.16$ in favor of men. The odds ratio thus ranks the disparity in the first case as greater than that in the second, which seems wrong, or at least counterintuitive.

An important property of the odds ratio is invariance: It remains unchanged if both the outcome (pass or fail) and antecedent (male or female) factors are inverted. For example, the ratio of the odds on passing for men to the odds on passing for women is equal to the odds on failing for women to the odds on failing for men. In the proficiency test illustration, the odds on passing for blacks to the odds on passing for whites are the same as the odds on failing for whites to the odds on failing for blacks; in both cases they are 0.585. Note that since the odds ratios are the same for pass and fail rates, the troubling inconsistency we noted would not have arisen if the 80% EEOC rule had been stated in terms of odds ratios rather than rate ratios.

In general, whenever the relative risk shows a disparity in proportions, the odds ratio shows a greater disparity in the same direction. Thus, when the relative risk is greater than one, the odds ratio is greater than the relative risk. (For example, in the 14–7% example the relative risk is 2 and the odds ratio is 2.16, in both cases in favor of men.) When the relative risk is less than one, the odds ratio is smaller than the relative risk. But perhaps most importantly, when rates for the groups being compared are both very low – as in epidemiological studies of rare diseases – the odds ratio is approximately equal to the relative risk, and is used as a surrogate for it in making causal inferences from epidemiological data. See Chapter 9.

A drawback of the odds ratio (and the relative risk) is that, because they deal with ratios of rates, they take no account of the absolute numbers involved. A Supreme Court case, *Craig v. Boren*,[29] illustrates this limitation. That case grew out of a 1958 Oklahoma statute that prohibited sale of "nonintoxicating" 3.2% beer to males under the age of 21 and to females under the age of 18. Statistical surveys, as interpreted by the Court, showed that approximately 2% of males in the 18–20 age range were arrested for driving while under the influence of alcohol, while only 0.18% of females in that age group were so arrested. Although the odds ratio and relative risk associated with these data were very high (in both cases more than 11), indicating a strong association between maleness and DWI arrests, Justice Brennan, writing for the Court, found the fit "tenuous" because of the very small percentages of males and females involved. Since the justification for discriminating against males was inadequate from this vantage point, the Court held the statute unconstitutional.

---

[29] 429 U.S. 190 (1976).

Chapter 3
Compound Events

In probability there are simple and compound events. Simple events are outcomes of random experiments that are not combinations of such outcomes. Compound events are combinations of simple events. To illustrate, suppose we toss a coin twice. The possible outcomes of this experiment are HT, TH, HH, and TT. Each of these results is a simple event with probability 1/4. Heads on the first toss may be viewed as a compound event consisting of the simple events HT and HH. The probability of *at least* one head is a compound event composed of the simple events HT, TH, and HH. Almost all events for which probabilities are computed in law are compound events, some of them quite complex combinations of simple events. To understand how to compute and interpret the probabilities of such compound events we need to understand the rules for combining their simple components. There are two fundamental rules for such combinations: the addition rule and the product rule. We discuss them and certain problems of interpretation below.

The Addition Rule

Suppose persons are selected at random for jury service from a large wheel that is 20% black and 10% Latino. What is the probability that the first person picked will be either a black or a Latino? Answer: Since these are mutually exclusive events (if you select a black you can't select a Latino, and vice versa) the probability of either one or the other occurring is the sum of their individual probabilities, or 0.20 + 0.10 = 0.30. This illustrates the fundamental addition rule of probability: *The probability of any one of a number of mutually exclusive events is equal to the sum of their individual probabilities.*

Summing probabilities of mutually exclusive events is a very common technique in applied probability calculations. We illustrate this with a famous case. In 1894, Alfred Dreyfus, a captain in the French General Staff, was convicted in a secret courts marshal of treason for betraying French military secrets to the Germans. The case turned on whether he was the author of a handwritten *bordereau* (note) that transmitted five memoranda purportedly containing secret military information to

M.O. Finkelstein, *Basic Concepts of Probability and Statistics in the Law*,
DOI 10.1007/b105519_3, © Springer Science+Business Media, LLC 2009

the German ambassador in Paris. Dreyfus was Jewish and anti-semitism was widely seen as having played a role in his conviction.

Dreyfus was incarcerated under unspeakable conditions in Devil's Island, the French prison colony. Back home, France was torn between Dreyfusards and anti-Dreyfusards. In 1899, public outcry forced a public retrial of the case. At the retrial, the famous criminologist Alphonse Bertillon appeared for the prosecution to analyze the handwriting of the bordereau. The handwriting did not look like Dreyfus's, but the army argued that Dreyfus contrived to make the handwriting look different to disguise his authorship. In support of that theory Bertillon testified that there were suspicious "coincidences" of the initial and final letters in 4 of the 13 polysyllabic words in the bordereau. Evaluating the probability of such a coincidence in a single word in normal writing as 0.2, Bertillon argued that the probability of four sets of coincidences was $0.2^4 = 0.0016$ in normal writing (using the product rule, which I discuss later). This low probability suggested to him that the handwriting in the document was not normal, which connected with the prosecution theory that Dreyfus had disguised his own handwriting to conceal his authorship. Dreyfus was again convicted.

Bertillon's premises remained spooky: The definition of coincidences and why they would make writing non-normal were never explained, and the probability of their occurrence in a single word was evidently pulled from the air. But even given his premises, Bertillon made the wrong calculation. His mistakes were later pointed out by an expert panel of the *Academie des Sciences* appointed by the court of appeals. Among Bertillon's mistakes was his failure to recognize that 4 coincidences in 13 polysyllabic words is a compound event consisting not only of the particular words in which they were found, but of every possible group of 4 of the 13 words. It is clear that Bertillon would have rejected the hypothesis of normal writing if any one of those groups had shown four coincidences, so the probability of coincidences in any of those groups would have to be included in the calculation. There are 715 such groups. If coincidences occur in only one group of four words, their occurrence in one group precludes their occurrence in another group, so the component events are mutually exclusive. Applying the addition rule, the probability of coincidences for each of the 715 groups of 4 words would have to be added to give the probability of such coincidences in *any* group of 4 words. Taking seriously Bertillon's evaluation that the probability of such a coincidence in a single word was 0.2, the probability of 4 coincidences in a particular group of 4 words and in none of the others is 0.000215 (we will show how this is computed when we discuss the product rule). The probability of exactly 4 coincidences in any 4 words is 0.000215 $\times$ 715 = 0.1535. By the usual standards, not a small probability.

Moreover, Bertillon claimed, in substance, that it was unlikely that normal writing would have contained as many as four coincidences. Since he would reject the hypothesis of normal writing for four coincidences, he must also do so if there were more. Thus he should have calculated the probability of there being 4, 5, ..., 13 words with coincidences. These events are mutually exclusive because if there are a given number of coincidences there cannot be more (or fewer). A separate calculation has to be made for each number of coincidences and, by a second application of the addition rule, these probabilities added for the probability of 4 *or more* coincidences in *any* of the 13 words. When the proper calculation is made, this probability is 0.2527. The *Academie des Sciences* conclusion: There was no basis for rejecting the null hypothesis that the writing was normal. The guilty verdict against Dreyfus was annulled and Dreyfus, having endured 5 years on Devil's Island, was restored to the army, promoted to major, and decorated with the *Legion of Honneur*. In 2006 there were memorial services in the *Chambre des Deputies* to honor the 100th anniversary of his exoneration.

If there are two component events that are not mutually exclusive, so that both could occur, the probability of the compound event is the sum of the probabilities of the nonexclusive events that comprise it, less the probability of their joint occurrence. To illustrate this using the coin-flipping example, suppose we flip a coin twice and ask for the probability of getting at least one head. By simple addition this would be the probability of heads on the first toss, which is the probability of the compound event HT + HH, plus the probability of heads on the second toss, which is the probability of the compound event TH + HH. The probability of each of these is one-half, so their sum is 1, which is clearly wrong since we may get two tails. The problem is that the events are not mutually exclusive since tossing heads the first time does not mean that we cannot toss heads the second time, and vice versa. Because of this overlap, by simple addition we have counted HH twice, once with each event. Subtracting the probability of one of the double countings gives us the correct answer: $0.50 + 0.50 - 0.25 = 0.75$.

If there are more than two component events, the adjustment for double counting is more complex. To cut through the complexity, a simple theorem eponymously called Bonferroni's inequality, is frequently used. It is this: *The probability of one or more occurrences of the simple events comprising a compound event is less than or equal to the sum of their probabilities*. If the probability of multiple occurrences is quite small, the sum of the probabilities of the simple events will be quite close to the probability of the compound event and is frequently used as an estimate for it.

To illustrate Bonferroni's inequality, consider a case in which an unusual carpet fiber is found on the body at a murder scene. It is compared with fibers found in the carpets associated with the suspect's apartment. Assume that if the crime-scene fiber did not come from the suspect's apartment there is 0.01 chance that it would match a fiber picked at random from one of his carpets. But if there are 50 distinguishable fibers in the suspect's carpets (each with a 0.01 chance of matching) the probability of one or more matches, if we try them all, is by Bonferroni's inequality, less than $0.01 \times 50 = 0.50$. It is in fact 0.395, somewhat less than 0.50 because the probability of more than a single match is not small. Thus a great deal turns on the way matches are tried and reported. The disparity between the probability of the observed match considered in isolation (0.01) and the probability of *some* match given multiple trials (0.395) is sometimes called *selection effect*.

The problem of selection effect is no mere academic nicety. The Federal Drug Administration requires drug companies in their tests of new drugs to identify in advance the group of patients who, the company believes, would benefit from it.

The reason is that many subgroups are possible and some subgroup is likely to show a statistically significant positive response to a drug even if it is ineffective. This is what happened to Centocor, the firm that promoted the Centoxin, a drug touted as highly effective for the treatment of a particular type of sepsis, gram-negative bacteremia. Before the study began, the company predicted that patients who took the drug would have a greater survival rate at 14 days than those who took a placebo. But during the clinical test, after an interim look, the company shifted the study endpoint to 28 days because the 14-day endpoint showed no improvement, while the 28-day endpoint showed a small improvement. Because of the change, the FDA rejected the study. Subsequent studies indicated that the drug was ineffective and it was ultimately withdrawn from further testing after an expenditure of almost a billion dollars. Note that the existence of selection effect does *not* mean that the probability of a particular result is changed. If we toss a coin a hundred times the probability of one or more heads is almost one, but the probability of heads on a particular toss is still one-half and is unaffected by the fact that we made other tosses.

In the Dreyfus case and our fiber example, the probability of at least one "hit" could be computed because we could count the number of trials. But in many cases this cannot be done. The cases arising out of the cluster of childhood leukemia in Woburn, Massachusetts, are illustrative of when counting can and cannot be done. Between 1969 and 1979 there were 12 cases of childhood leukemia in Woburn, when only 5.3 were expected based on national rates. Two municipal wells, G and H in Fig. 3.1, were found to be contaminated and the culprit appeared to be industrial waste from two nearby dump sites. Two companies were implicated in dumping waste at those sites. Lawsuits were brought against the companies on the theory that the contaminated well water had caused the leukemia. The story of the cases is told in Jonathan Harr's best-selling book, *A Civil Action*, from which a movie of the same name was made, starring John Travolta as the crusading lawyer for the plaintiffs.

There is less than a 1% probability of such a large number of childhood leukemia cases in Woburn, given national rates, and so by the usual 5% level of statistical significance we would reject the hypothesis that the long-run rate of leukemia in Woburn was no greater than national rates. There are, however, an uncountable number of geographic entities in which a cluster of leukemia might arise and receive notice. With many such possibilities some clusters are to be expected, even if the risk were everywhere the same. If we saw nefarious influences at work whenever there was such a cluster, the rate of false accusations would be unacceptably high.

On the other hand, the clustering of cases *within* Woburn is a more defined problem for which an adjustment can be made for multiple trials. There were six approximately equal census tracts in Woburn. Six leukemia cases were clustered in tract 3334, which was adjacent to the tract in which the wells were located. See Fig. 3.1. Adopting the hypothesis that a leukemia case is equally likely to appear in any tract because the well water was not responsible, the probability of six or more cases in any single tract, given that 12 cases occur, is 0.0079. By Bonferroni's inequality, the probability of six or more cases in at least one of the six tracts is less than six times

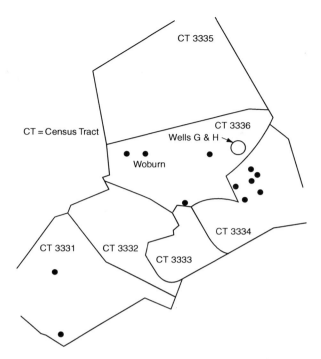

Fig. 3.1 Residences of childhood leukemia patients. Woburn, Mass, 1969–1979

the probability of that event in a single tract, or $6 \times 0.0079 = 0.047$. The exact probability of six or more cases in any tract is slightly less than this sum because there is a minuscule probability (0.000006) of six leukemia cases in two tracts, i.e., the component events are not mutually exclusive, although nearly so. With this result one would be justified (although marginally) in rejecting the hypothesis that a case was equally likely to appear in any tract.

A stronger result is possible. In the Dreyfus case and in the fiber example we had to consider all possible outcomes because the event of interest was whether *any* four words had coincidences or *any* of the fibers matched; which words had coincidences or which fibers matched was irrelevant. In the Woburn case that is no longer true. The fact that the excess leukemia was found in the census tract adjoining the wells and not in a more remote tract is a point of evidence in favor of causation. Although the town's pipes were interconnected, it appears that more of the water from contaminated wells went to nearby houses. Assuming that propinquity to the wells was a substantial a priori reason to focus on the tracts in which the leukemia had appeared, the event in question could have been viewed as the probability of an excess of leukemia in those tracts alone and summing probabilities over other tracts would have been unnecessary. This restriction could validly be imposed only if we would not have rejected the null hypothesis if the excess had appeared in a more remote tract instead of one proximate to the wells. On that understanding, the null hypothesis is still that leukemia was equally likely to occur in any tract, but the event whose probability was being computed would be the observed excess of leukemia in the two or three nearby tracts, so that the probability of the observed excess in the particular tract in which it occurred would be multiplied by two or three instead of six. This would give a higher level of significance.

While an a priori basis for focusing on what has in fact been observed need not be discovered prior to knowing the data, but need only be a priori in the sense of being independent of them, reasons uncovered post hoc must be scrutinized with care. One has to be cautious about such arguments because, with a little imagination, it may be quite easy to come up with post hoc theories to explain an observed cluster of disease and thus seemingly to justify computing the probability of the cluster as if it had been selected on some a priori basis. The Woburn case has a whiff of this problem because the scientific underpinning for the theory that contaminated water causes leukemia is weak; no one really knows what causes leukemia.

On the other hand, it is not always appropriate to adjust for selection effect merely because multiple trials are made in a search for a match. If the hypothesis being tested relates to the specific match that has been made on one of the trials, adjustment would be incorrect because, as we have observed, the probability of a match on that trial is unaffected by other trials. The calculation of probabilities in DNA identification cases provides an important example of some confusion on this point. In a number of criminal investigations suspects have been initially identified by computerized searches of DNA databases. As the number and size of such databases increase, it is likely that initial identifications will more frequently be made on this basis. In a report, the Committee on DNA evidence of the National Research Council stated that in such cases the usual calculation of match probabilities had to be modified. As one of two modifications, the Committee recommended that the calculated match probability be multiplied by the size of the database searched.[1] This is obviously an application of Bonferroni's inequality to reflect the search. The adjustment in this context is wrong, however. What is relevant is not the probability of finding *a* match in the database, but the probability of finding *the observed* match, given that the suspect is innocent; that probability is described by the usual match probability calculation. How many others were tested is essentially irrelevant.

Our position that no adjustment in probability calculation for multiple trials is required when multiple suspects are tested, one test for each, may seem somewhat paradoxical. After all, if one makes many trials so that finding at least one match by chance becomes likely, why should the match that is found be strong evidence against an accused? How is it that something we are likely to find by coincidence in a screening search may be strong evidence against coincidence when we find it?

To see that there is no paradox, let us return to the unusual carpet fiber found on the body at a murder scene. Suppose that the unusual fiber occurs with a frequency of 1 in 500 carpets. Suppose that the crime has been committed in a town that has

---

[1] Nat'l Research Council, "The Evaluation of Forensic DNA Evidence," 32 (1996).

5,000 homes, each with a carpet, and the guilty party is believed to come from one of them. A large-scale screening search is made by taking a fiber from each house and comparing it with the unusual fiber found at the crime scene. Since the rate of such unusual fibers is 1 in 500 we would expect to find 10 homes with such fibers.[2] Putting aside the effect of other evidence, the probability that the fiber came from the suspect's home (which has a matching fiber) rises from about 1 in 5,000 before the identification to about 1 in 10 afterward, i.e., by a factor of 500. The same factor of 500 would apply regardless of the size of the suspect population or if, as is usual, the size of that population is unknown. It is also unchanged if the search is less than exhaustive and the suspect's home is the only one or only one of a few homes identified as having such a fiber. The hypothetical illustrates that the probative force of an evidentiary trace represented by the factor of 500 comes from narrowing the suspect to a small group out of a larger population; the fact that we're almost certain to find at least one member of the group from a screening search is irrelevant to the probative effect of the reduction in number of suspected sources.

In the DNA case, the probabilities associated with matching by coincidence are so small relative to the size of the databases searched that adjustment for multiple comparisons may not make much difference to the result. But in other cases it can matter a great deal. We must defer discussion of such a case until the next chapter.

The Product Rule

Sylvia Ann Howland died in 1865 in the old whaling town of New Bedford, Massachusetts. She left a will bequeathing the income from about half of her very large estate to her niece, Heddy B. Robinson, who lived with her. After the probate of this will, Heddy Robinson brought a suit seeking performance of an agreement between them to make mutual wills leaving their property to each other. In support of that claim, Ms. Robinson produced an earlier will of her aunt leaving all her property to Robinson. This will contained a curious second page that was separately signed, purportedly by the deceased, reciting the agreement to leave mutual wills. There was a second copy of this separate page, also purportedly signed by the deceased. Robinson said that her aunt signed the copies of the second page at the time she signed the will. The executor rejected Robinson's claim of agreement on the ground that the signatures on the second page appeared to be tracings from the undisputed signature on the will. Robinson sued and the case became a cause célèbre.

Among the parade of celebrated witnesses (which included Oliver Wendell Holmes, Sr.) was a noted Harvard professor of mathematics, Benjamin Peirce. He was assisted by his son. Charles Sanders Peirce, who later became a famous philosopher. In what appears to be the first use of statistical inference in a law case, Peirce

---

[2] The probability of finding at least one match by chance, none of the houses in the town being the source, is $1 - (499/500)^{5,000} = 0.99997$, or more than 99.99%.

père set out to calculate the probability that the agreement between the signatures would have been so close if both had been genuine.

Peirce said that there were 30 downstrokes in Sylvia Ann Howland's signature. He examined these in the authentic and one of the disputed signatures and found that all of them "coincided." In using such coincidences he anticipated modem handwriting analysis. He then counted the number of such coincidences in every possible pairing of 42 undisputed signatures of Sylvia Ann Howland taken from various other documents. Since there were 861 pairs and 30 downstrokes for each pair there were 25,830 comparisons, of which 5,325, or slightly over 1 in 5, corresponded. Using 1/5 as the probability of two downstrokes coinciding, the probability of 30 such coincidences he calculated as that figure to the 30th power, which he testified (making a mistake in arithmetic), was "once in 2,666 millions of millions of millions." (The correct answer is once in 379×10^{18} times if the 5,325/25,835 figure is used and once in 931×10^{18} times if 1/5 is used.) The result was mind-boggling and it boggled Peirce. He testified, "This number far transcends human experience. So vast an improbability is practically an impossibility. Such evanescent shadows of probability cannot belong to actual life. They are unimaginably less than those least things which the law cares not for....The coincidence which is presented in the case therefore cannot be reasonably regarded as having occurred in the ordinary course of signing a name."[3]

The assumption of this calculation is the so-called independence theorem (or definition) of elementary probability. The probability of the joint occurrence of events A and B is the probability of A given B times the probability of B, or equivalently, the probability of B given A times the probability of A. If A and B are independent events, then the probability of A is unaffected by whether or not B has occurred; this is simply the probability of A unconditional on B. *Thus in the special case in which events are independent, the probability of their joint occurrence is equal to the product of their individual probabilities.* For example, in the Dreyfus case, assuming that a coincidence in one word is independent of a coincidence in another, Bertillon should have computed the probability of coincidences in the particular four words and in none of the remaining nine as the product of the probabilities of those 13 events, to wit, $(0.2)^4 \times (0.8)^9 = 0.000215$.

The product rule rather rapidly generates extremely small probabilities, but it is frequently questionable whether the events whose probabilities are multiplied are in fact independent. In the Howland Will case, in multiplying probabilities together, Peirce assumed that whether one pair of downstrokes coincided was independent of whether another did in the same pair of signatures. Apparently, he was not chal-

---

[3] The Howland Will Case, 4 Am. L. Rev. 625, 649 (1870). See also Louis Menand, "The Metaphysical Club," 163–176 (2001) (describing the case in detail). The circuit court never ruled on whether the signature was a forgery. Instead, it held that Hetty's testimony on her own behalf during the trial violated a statute prohibiting parties to a suit from giving testimony unless called by the other side or commanded to testify by the court. Since Hetty was the only witness to her contention that her and her aunt's wills had been mutual, "the court is of the opinion that the contract is not proved." Hetty lost, but was still a rich woman, and went on to become a phenomenally successful moneylender on Wall Street.

lenged on this ground, but he should have been. For almost certainly it was not true. A priori, signatures that were written at about the same time and under similar conditions (as the genuine and disputed signatures purportedly were) are more likely to match than those written at some time apart, or under different conditions (as the signatures were that were used in the data). Therefore, coincidences would tend to come not single spy but in battalions; encountering some would increase the probability of finding others. A more detailed modern analysis of Peirce's data shows that in fact there are far too many pairs of signatures with high and low numbers of coincidences to be consistent with the independence model.

A similar misuse of the product rule was made years later in the *Risley* Case previously discussed.[4] The prosecution expert assumed that the probability of a defect in a typewritten letter was some unspecified number; multiplying these component probabilities together, he concluded that the joint probability of all 11 defects was about one in four billion. We have no data, but very probably defects were not independent events since their occurrence would be related to the age of the machine. The New York Court of Appeals correctly objected that the testimony "was not based upon observed data, but was simply speculative, and an attempt to make inferences deduced from a general theory in no way connected with the matter under consideration supply the usual method of proof."[5]

Our third example is the famous *People v. Collins*,[6] which involved a frivolous use of the independence model that has nevertheless captured the imagination of a generation of evidence scholars. A woman was assaulted from behind and robbed in an alleyway. She managed to see that her assailant was a blond woman. A witness saw a blond woman run out of the alleyway and enter a yellow car driven by a black male with a beard and a mustache. A few days later, following a tip, officers investigating the crime arrested a couple answering this description. At the trial, to bolster identification, the prosecutor called an instructor in mathematics and asked him *to assume* the independence and frequency of the following factors: yellow automobile (1/10); man with mustache (1/4); girl with ponytail (1/10); girl with blond hair (1/3); black man with beard (1/10); interracial couple in car (1/1,000). Multiplying these rates together, the witness concluded that the rate of such couples in the population was 1 in 12 million. The defendants were convicted.

On appeal, the Supreme Court of California reversed, pointing out that there was no support for the claimed frequencies or their independence. That there was no support for the claimed frequencies was clearly right; the assumption was a pure speculation. On the lack of independence, the court was right to conclude that the factors were not independent (among other things, interracial couple is correlated with black man with beard and girl with blond hair) but possibly wrong to conclude that the probability of the joint occurrence of the factors was not the product of their individual probabilities. After all, *if* the probabilities were conditional, multiplying them together would have been appropriate, without independence. Thus, if among

---

[4] See p. 3.

[5] *Id.*, 108 N.E. at 203.

[6] 68 Cal. 2d 319 (1968) (en banc).

interracial couples in cars there was a black man with a beard in 1/10 of them, then the joint occurrence of interracial couple and black man with beard is indeed 1/1,000 × 1/10. And if among those interracial couples in cars in which there is a bearded black man, the rate of girls with blond hair is 1/3, then the joint occurrence of all three events is 1/1,000 × 1/10 × 1/3, etc. Of course, the exercise is entirely fanciful since the component probabilities were hypothetical.

Here is a more serious recent matter in which the court rejected the assumption of independence. In 1999 Sally Clark was convicted, in England, for the murder of her baby sons, Christopher and Harry. Christopher was born in 1996 and Harry about 1 year later. Both were apparently healthy babies who had died, one at 8 weeks and the other at 11 weeks of age. The cause of death was unknown; hence the diagnosis after the first death was Sudden Infant Death Syndrome (SIDS). After the second infant died, the mother was arrested and charged with murder in both their deaths. At Mrs. Clark's trial an eminent pediatrician, Professor Sir Roy Meadow, testified to a large government-funded study of deaths of babies in five regions of England between 1993 and 1996, which was just being released. The study found the risk of a SIDS death as 1 in 1,303. But if the infant was from a non-smoking family that had a wage earner, and the mother was over age 26 or the infant was an only child, then the chance fell to 1 in 8,543. Since the Clark family had those features (both parents were solicitors), the 1:8,543 figure was applicable. The report also calculated that the risk of two such deaths in such a family was 1 in 8,543 × 8,543 = 1 in 73 million. Professor Meadow testified to that figure – which made headlines – but did not give the study's qualification that the figure did not "take account of possible familial incidence of factors" other than those in the study and that in such dual-death SIDS cases it would be "inappropriate to assume that maltreatment was always the cause." Since there were 700,000 live births per year in the United Kingdom, Professor Meadow concluded in his testimony that "by chance that happening [two SIDS deaths in a family] will occur about once in every hundred years." He also testified to parts of the postmortem findings, which he believed suggested a non-natural death in both cases, apart from the statistics.

No objection was made to this testimony at trial, but on rebuttal an expert for the defense, a Professor Berry, one of the editors of the SIDS study, challenged the 1:73 million figure on the ground of familial factors. He made the point that simply squaring the SIDS rate was an oversimplification because it did not take into account that the two deaths could have been caused by a common familial factor. In summing up the evidence, the trial judge cautioned the jury not to put much reliance on the statistics, commenting that "we do not convict people in these courts on statistics," and noting that "If there is one SIDS death in a family, it does not mean that there cannot be another one in the same family." However, the jury, by a 10-2 vote, convicted the mother.

On the first appeal the court dismissed objections to the 1:73 million figure. The court pointed out that the arguments against squaring were known to the jury and the trial judge repeated them in his summing up. In addition, the prosecution was only making the broad point that two SIDS deaths in a family were very unusual, and in that exercise the precise figure was unimportant.

On a second appeal, the verdict was challenged on the ground of newly discovered medical evidence indicating that Harry had died of an infection and on the ground of misleading statistics. On the second day of the argument, the prosecution announced that it would not longer defend the verdict. The court then reversed the conviction on the ground of the new evidence. In dictum, the court said that it would also have reversed on the statistics. The court described Meadow's evidence as "manifestly wrong" and "grossly misleading," commenting that there was evidence to suggest that two SIDS cases in a family was "a much more frequent occurrence" than was indicated by the 1 in 73 million figure. From the data of the study this seems correct: Of the 323 families with a SIDS baby in the study 5 of them had a prior SIDS baby; of the 1,288 control baby families (there were four contemporaneous babies used as controls for each SIDS case) there were only 2 prior SIDS cases. Thus the odds of a SIDS baby were about 10 times greater for a family with a prior SIDS case than for a family without such a case.

In a perhaps unprecedented development, after the case was closed, Sally Clark's father filed a complaint with the General Medical Council against Professor Meadow for unprofessional conduct arising from his misleading evidence in the case. After a hearing, the Council found that his conduct as a witness had been "fundamentally unacceptable" and struck him from the registry of physicians. He was, however, later reinstated.

DNA testing is another highly important context in which the independence assumption plays a key role. At certain locations on a strand of DNA, the sequence of bases that forms the DNA molecule consists of a number of short repeating patterns, called alleles, that vary from person to person. The alleles at the same location for the two chromosomes in a pair together are called a genotype. If the two alleles are the same, the genotype is called homozygous; if they do not match, the genotype is heterozygous. Genotypes of two different individuals match if both alleles match. The variety of genotypes that can be distinguished at a location is quite large. For example, at one particular location, according to FBI data, blacks have at least 909 genotypes, whites have at least 901, etc. Moreover, the same test can be applied to different locations on the DNA strand.

Forensic laboratories routinely multiply together the frequencies of alleles on each strand of DNA and at the various locations to determine the joint frequency of all of the observed alleles. In the FBI database, which at one point had fewer than 1,000 individuals for each of four racial groups, the probability that a genotype from two different individuals would match at a particular locus ranges from about 0.03 to 0.001. What gives the method its power is the multiplying together of such probabilities, which yields joint frequencies of less than one in millions or billions. The assumption that the paternal and maternal strands of DNA are independent is called Hardy-Weinberg equilibrium and the assumption that alleles at different locations are independent is called linkage equilibrium.

When experts from DNA testing laboratories began testifying in court, these assumptions were challenged as speculative by commentators, who pointed out that Hardy–Weinberg equilibrium would be violated if there were subgroups in the population that had markedly different frequencies of alleles and tended to intermarry.

In that case, observing an allele common to the subgroup that was contributed by one parent would be associated with seeing a similar allele contributed by the other parent, with the result that the frequency of homozygous genotypes would be greater than one would expect under independence. Critics of DNA testing reported observing an excess of homozygotes, suggesting a violation of independence. Supporters of testing replied that the observed excess was merely the inability of the tests to distinguish alleles that were not very different in weight.

As for linkage equilibrium, detailed analysis of databases showed that the observed joint frequencies of pairs and triplets of alleles were consistent with what one would expect under independence.[7] This evidence is supportive of the independence hypothesis, but does not go so far as to justify the claimed frequencies of matches on the order of 1 in millions or even billions claimed by enthusiastic experts. In the U.K. case of *Regina v. Alan Doheny*,[8] the Crown's expert came up with a random occurrence ratio of 1 in 40 million for the DNA profile in the case, but the defense objected that he had improperly assumed independence of a multi- and a single-locus probe in multiplying together their respective frequencies. The Crown's experts conceded that a dependence could arise if the multi-locus probe identified bands that were the same as or in close proximity to those identified by the single-locus probe. The appeals court agreed with the experts that there was a risk of dependence that could not be quantified, analyzed the evidence using the much higher random occurrence ratio associated with only the single-locus probe, and quashed the conviction.

In fact, even if there is true independence the fantastically small random occurrence ratios generated in the testimony can nevertheless be quite misleading because the number that reflects the import of the evidence is not the frequency of true matches, but the frequency of called matches, and this includes true matches and mistakes. The latter often is ignored because it is much harder to quantify given that proficiency testing has been quite limited. Mistakes include things like accidentally switching samples to declaring a match despite discrepancies because in the opinion of the technician the failure to match is due to some artifact of the test rather than a difference in the DNA. For example, the sample found at the crime scene may show an extra band that the defendant lacks, but the technician may conclude, judgmentally, that it is caused by contamination of the sample and declare a match. Considerations of this sort cast doubt on the probative value of the intimidating numbers generated by the independence model, but the assumption that coincidental matching of DNA among unrelated individuals is very rare is now widely accepted.

---

[7] If the sample is large enough the joint frequency of pairs or even triplets of attributes can be directly observed to determine whether their number is consistent with the hypothesis of independence. If so, a violation of independence would then require correlation in higher orders of attributes. This would seem unlikely since the physical mechanism for producing attributes is such that if there were correlations among higher orders there should be evidence of it in pairs or triplets. Thus as DNA databases grow in size, tests can be performed to lend plausibility to the independence hypothesis.

[8] [1996] EWCA Crim. 728 (July 31, 1996).

Chapter 4
Significance

Women employees bring a gender discrimination class action against their employer, citing the fact that their average salary is $2,000 less than that of men in the same division of the company. A labor economist, appearing for the women, testifies that the difference in averages between men and women is statistically significant. In another case, a commodities broker with discretionary authority over accounts F and G is accused by account F owners of siphoning off profitable trades to account G, in which the broker had an interest. An expert for the broker testifies that the lower rate of profitable trades in account F is not statistically significant.

The testimony regarding statistical significance in these cases seems relevant to the issues in the actions, but what exactly does it mean? And, in particular, what does it imply for the weight of the evidence that there was gender discrimination, or that the broker had siphoned off trades?

The Concept of Significance

The concept of significance is fundamental to statistical inference but it has a narrower and more technical meaning in statistics than it does in common parlance or in law. The technical definition begins with the idea that the data on which the experts based their statements are in each case random samples from some larger universe or population. It is further assumed that in this larger population there is no difference in the data for the two groups; the apparent differences in each case are due to the accidents of sampling that would in the long run even out. These assumptions embody what is called a null hypothesis. Assuming that the null hypothesis is true, it is possible to compute the probability of observing such a large disparity as that appearing in the sample. This probability is called the *attained level of significance* or *P-value*, or sometimes the *tail probability*. The observed disparity is said to be significant when the *P*-value is quite small. When the *P*-value is small, most statisticians would reject the null hypothesis, concluding in the employment case that something more than mere chance is at work. Conversely, in the broker case, since the difference was nonsignificant, statisticians would conclude that chance alone cannot be ruled out.

M.O. Finkelstein, *Basic Concepts of Probability and Statistics in the Law*,
DOI 10.1007/b105519_4, © Springer Science+Business Media, LLC 2009

How small is "quite small"? The conventional choice is 0.05, or sometimes 0.01. Using the first of these choices, if the probability of observing a difference as large as, or larger than, that observed, given the null hypothesis, is less than 5%, the difference is deemed statistically significant. The effect of this rule is that in cases in which the null hypothesis is true we are willing to make mistakes 5% of the time in rejecting it. Errors of rejecting the null hypothesis when it is true are called Type I errors. Errors of failing to reject the null hypothesis when it is false are called Type II errors. The rate of Type I errors describes the level of significance; the rate of Type II errors relates to power, which is discussed in Chapter 7. The conventional nomenclature is that the smaller the P-value the higher the level of significance.

There has been some unease in the legal community about the arbitrariness of applying the statistical mantra of 5 or 1% significance levels. As a practical matter, however, other issues that arise in assessing significance are usually far more important than the alternative significance cutoffs commonly suggested (e.g., 10%). Some commentators, however, would go much further; they argue that 5% is an arbitrary statistical convention and since preponderance of the evidence means 51% probability, lawyers should not use 5% as the level of statistical significance, but 49% – thus rejecting the null hypothesis when there is up to a 49% chance that it is true. In their view, to use a 5% standard of significance would impermissibly raise the preponderance of evidence standard in civil trials.

Of course the 5% figure is arbitrary (although widely accepted in statistics) but the argument is fallacious. It assumes that 5% (or 49% for that matter) is the probability that the null hypothesis is true. The 5% level of significance is not that, but the probability of the sample evidence *if* the null hypothesis were true. This is a very different matter. As I pointed out in Chapter 1, the probability of the sample given the null hypothesis is not generally the same as the probability of the null hypothesis given the sample. To relate the level of significance to the probability of the null hypothesis would require an application of Bayes's theorem and the assumption of a prior probability distribution. However, the courts have usually accepted the statistical standard, although with some justifiable reservations when the P-value is only slightly above the 5% cutoff. I discuss one such case later in this chapter.

The notion that what the world shows us is a random selection from some larger universe or population may be quite explicit or more hypothetical. Political polling in which 1,000 likely voters are selected by random digit telephone dialing and the results extrapolated to the entire electorate is an example of an explicit larger population (the electorate) from which a sample has been selected by a random process. On the other hand, in a sex discrimination law suit, where the figure of interest is the difference in average salaries between men and women, the entire population of employees at a given time is not a sample in the same concrete sense. But the particular observed disparity may still be thought to reflect transient, random factors. Usually, the employees at a given time are thought of as a sample from the stream of employees over time. Alternatively, the particular salary assignments to men and women may be viewed as only one realization out of all possible assignments of those salaries to the employees. We can model that latter conception by assuming that the salary figures for the employees are written on chips and put in an urn. We pick from the urn, at random, a number of chips equal to the number of men; those are the male salaries from which we calculate the male average; the chips remaining in the urn are the female salaries from which we calculate the female average. This is done repeatedly and the average salaries of the two groups after each selection are recorded. If in fewer than 5% of the selections the difference between men and women equals or exceeds $2,000 the difference is said to be significant at that level. Thus even when no express sampling is involved, the data may still be viewed as one random realization among many possibilities.

In the commodity account example, the total number of transactions in both accounts can also be represented as chips in an urn. The profitable chips are so marked. A number of chips are then withdrawn at random from the urn, the withdrawn chips representing those made in account F; those remaining are in account G. The exercise is repeated many times and if the number of profitable trades in G equals or exceeds the number observed in G at least 5% of the time, G's advantage is deemed nonsignificant, and we cannot reject the possibility that a difference that large between F and G in rate of profitable trades was caused by chance.

The definition of "such a large disparity" has two aspects. First, it means a disparity as large or larger than the one observed in the sample. Such probabilities are sometimes called *tail probabilities* because they include all the extreme values in the tail of a probability distribution, namely, those that are highly improbable. They are to be distinguished from the probability of any particular disparity, which is called its *point probability*. Why include "or larger"? First, note that the point probability of any particular disparity – including even a zero disparity in exact agreement with the null hypothesis – is characteristically very small under the null hypothesis; it is the aggregate probability of the disparities at least as large as the one observed that describes the attained level of significance. The probability of tossing exactly 50 heads out of 100 trials with a fair coin – the expected number and the most likely single result – is still only about 8%; but the probability of tossing 50 *or fewer* heads is about $\frac{1}{2}$. And if we would reject the null hypothesis because the observed number in a sample was too small, we would also reject the hypothesis if it were even smaller; the tail probability thus gives us the total error for our rejection rule.

Second, "such a large disparity" may or may not include a disparity in the direction opposite to that appearing in the sample. Thus if a sample disparity in average pay is $2,000 in favor of men, the concept of "such a large disparity" could be defined as men earning on average at least $2,000 more than women, or as either group earning at least $2,000 more than the other. The test of significance is called one-tailed in the first case because the 5% level of significance relates to the probability of extreme values in one direction only (e.g., men earning more than women), and two-tailed in the second case because the 5% level of significance relates to the aggregate probability of extreme values in either direction (either group earning on average at least $2,000 more than the other group). I will return to this subject later in this chapter.

Rejecting the Null Hypothesis

Rejecting the null hypothesis in effect rules out chance as an explanation of the observed disparities, but not anything else. The rejection of chance is more important in a case in which chance is an explicitly or implicitly tendered issue than it is when other mechanisms may account for the observed difference. In a promotion case, chance is only involved at the level that factors favorable to promotion may be said to be randomly distributed between the two sexes. This is frequently disputed. For example, in *Ste. Marie v. Eastern R. Ass'n*,[1] Judge Friendly held, in a gender discrimination class action, that although the low numbers of women promoted to technical or managerial positions could not have been the result of chance, the disparity was not evidence of discrimination because the group from which the promotions were made included secretaries (overwhelmingly women) who lacked the training needed for promotion.

On the other hand, a difference that is not statistically significant means that chance *may* explain the difference, not that it necessarily does. If the sample is too small, a real difference may be declared nonsignificant only because the evidence is insufficient. We therefore do not say that a nonsignificant difference leads us to accept the null hypothesis, but only that we cannot reject it. Of course there may be other evidence indicating that chance is not the culprit. In *Waisome v. Port Authority of New York & New Jersey*,[2] a promotion case, there was such evidence. In that case, 14% of white officers were promoted but only 7.9% of black officers. Because only five blacks were promoted, the difference was not statistically significant. The court of appeals held, however, that the lack of significance did not matter because the evidence showed that the difference was not caused by chance, but by a written test, which caused blacks to be ranked lower than whites on an eligibility list from which the promotions were made.

The Two-or-Three Standard Error Rule

In calculating significance levels, a common procedure is to compute the number of standard errors between the observed sample data and the expected value under the null hypothesis. Counting standard errors (usually described as standard deviations) has been crystallized as the method for testing significance in federal discrimination jurisprudence. In *Castaneda v. Partida*,[3] a jury discrimination case involving underrepresentation of Mexican-Americans on grand juries, the U.S. Supreme Court observed in a footnote that "If the difference between the expected value and the observed number is greater than two or three standard deviations, then the hypothesis that the jury drawing was random would be suspect to a social scientist." There is

---

[1] 650 F.2d 395 (2d Cir. 1981).

[2] 948 F.2d 1370 (2d Cir. 1991).

[3] 430 U.S. 482, 496, n.17 (1977).

of course a non-trivial difference between two and three standard errors in terms of significance probability, but the Court did not have to be very precise about where to draw the line because the difference in that case was 29 standard errors! The lower courts, however, picked up this almost casual observation about social science and hammered it into a rule of law: Nothing less than two standard errors will do for a prima facie case.

If intended as a rule for sufficiency of evidence in a lawsuit, the Court's translation of social science requirements was imperfect. The mistranslation relates to the issue of two-tailed vs. one-tailed tests. In most social science pursuits investigators recommend two-tailed tests. For example, in a sociological study of the wages of men and women the question may be whether their earnings are the same or different. Although we might have a priori reasons for thinking that men would earn more than women, a departure from equality in either direction would count as evidence against the null hypothesis; thus we should use a two-tailed test. Under a two-tailed test, 1.96 standard errors is associated with a 5% level of significance, which is the convention. Under a one-tailed test, the same level of significance is 1.64 standard errors. Hence if a one-tailed test is appropriate, the conventional cutoff would be 1.64 standard errors instead of 1.96.

In the social science arena a one-tailed test would be justified only if we had very strong reasons for believing that men did not earn less than women. But in most settings such a prejudgment has seemed improper to investigators in scientific or academic pursuits; and so they generally recommend two-tailed tests. The setting of a discrimination lawsuit is different, however. There, unless the men also sue, we do not care whether women earn the same or more than men; in either case the lawsuit on their behalf is correctly dismissed. Errors occur only in rejecting the null hypothesis when men do not earn more than women; the rate of such errors is controlled by one-tailed test. Thus when women earn at least as much as men, a 5% one-tailed test in a discrimination case with the cutoff at 1.64 standard deviations has the same 5% rate of errors as the academic study with a cutoff at 1.96 standard errors. The advantage of the one-tailed test in the judicial dispute is that by making it easier to reject the null hypothesis one makes fewer errors of failing to reject it when it is false.

The difference between one-tailed and two-tailed tests was of some consequence in *Hazelwood School District v. United States*,[4] a case involving charges of discrimination against blacks in the hiring of teachers for a suburban school district. A majority of the Supreme Court found that the case turned on whether teachers in the city of St. Louis, who were predominantly black, had to be included in the hiring pool and remanded for a determination of that issue. The majority based that conclusion on the fact that, using a two-tailed test and a hiring pool that excluded St. Louis teachers, the underrepresentation of black hires was less than two standard errors from expectation, but if St. Louis teachers were included, the disparity was greater than five standard errors. Justice Stevens, in dissent, used a one-tailed test,

---

[4] 433 U.S. 299 (1977).

found that the underrepresentation was statistically significant at the 5% level without including the St. Louis teachers, and concluded that a remand was unnecessary because discrimination was proved with either pool. From our point of view. Justice Stevens was right to use a one-tailed test and the remand was unnecessary.

The two-standard-error rule is a way of estimating probabilities. It assumes that the underlying probabilities can be approximated by the normal distribution. However, the rule is probably better expressed in terms of probabilities because in some cases the normal distribution is not a good approximation of the underlying probabilities. The two-standard-error rule is also irrelevant where the hypothesis of random selection is questioned because the samples are *too close* to expectation rather than too far. This has occurred in the jury selection cases in which the data suggested the imposition of a quota. For example, in one case arising in Orleans Parish, Louisiana,[5] nine grand juries were selected over a 4-year period from 1958 to 1962. Each grand jury had 12 members. There were two blacks on eight of the juries and one on the ninth. Blacks constituted one-third of the adult population of the Parish, but the Supreme Court of Louisiana held that the low level of black representation was due to lower literacy and requests for excuse based on economic hardship. But even if it is assumed that the overall proportion of blacks actually serving ($17/108 = 15.7\%$) matched their proportion in the eligible population, so that we would expect a little fewer than two blacks per jury, the probability that actual representation would so closely mirror the expected numbers is only about 0.001. The null hypothesis of random selection must be rejected; most likely, a quota was at work.

In another case, a debt collection company was paid a percentage of the amount it collected on receivables it took on consignment. Based on experience, the company claimed that it would collect 70% of the receivables and reported its earned income on that basis before the receivables had been collected. The U.S. Securities and Exchange Commission began an investigation. To test the company's assertion, the accountant for the company said that he had taken three random samples of 40 receivables each, which showed collection rates of 69, 70, and 71%, respectively. However, assuming the 70% figure was correct, the probability that three random samples would be so close to the expected value of 70% is less than 0.05; hence the truth of the assertion that the three samples were selected at random, or that they yielded those results, would have to be challenged. The accountant's report was not credible and, indeed, he later confessed to having participated in the company's fraud.

Affirmative action plans are perhaps the most important arena where the existence of a quota is a key issue. In the landmark *Grutter v. Bollinger*,[6] the Supreme Court held that Michigan Law School's affirmative action plan for admission of minority students passed constitutional muster. The standard enunciated by the Court was that "universities cannot establish quotas for members of certain racial

---

[5] State v. Barksdale, 247 La. 198 (1964), *cert. denied*, 382 U.S. 921 (1965).
[6] 539 U.S. 306 (2003).

groups or put members of those groups on separate admission tracks.... Universities can, however, consider race or ethnicity more flexibly as a 'plus' factor in the context of individualized consideration of each and every applicant." Since, as the majority found, Michigan's program met these requirements it did not violate the Equal Protection clause. Four justices dissented.

The majority and dissenting justices agreed that admission rates for minorities that had too little variation were evidence of a forbidden quota, but disagreed over what statistics were relevant and what could be inferred from them. The majority focused on the number of minorities in each class of the law school and pointed out that between 1993 and 2000 this varied from 13.5 to 20.1%, a range it found "inconsistent with a quota." The four dissenting justices thought that the more relevant statistic was the proportion of minorities admitted to the law school, rather than those accepting. In this the minority would appear to be right. Looking at that data, Chief Justice Rehnquist, writing a dissenting opinion, thought that between 1995 and 2000 the "correlation between the percentage of the Law School's pool of applicants who are members of the three minority groups [African-Americans, Hispanics, and Native-Americans] and the percentage of the admitted applicants is far too precise to be dismissed as merely the result of the school 'paying some attention to the numbers.'" Undeniably, there was a close correspondence. Rehnquist included tables which showed the percentages and highlighted two examples: In 1995 when 9.7% of the applicant pool was African-American, 9.4% of the admitted class was African-American. In 2000, the corresponding numbers were 7.5 and 7.3%. The figures for other years and for other groups were almost as close, and in some cases closer. Rehnquist concluded that the numbers spoke for themselves: The "tight correlation between the percentage of applicants and admittees of a given race, therefore, must result from careful race based planning by the Law School."[7]

But do the numbers really imply race-based selection? The statistical case is not that unequivocal. If there had been a quota based on the racial proportions of applicants, the number admitted would have been closer to expectation than would have occurred by chance as a result of race-blind selection at random from the pool of applicants. No statistical analysis of that question was presented by the lawyers in the case (or by the justices). In one approach, if one computes the disparity between the numbers of admittees and their expected numbers based on the percentages of applicants for the four groups – African-Americans, Hispanics, Native-Americans, and Others (presumably whites, Asians, etc.) – for each of the 6 years and then combines the evidence across the 6-year period, the probability of such a "tight correlation" between observed and expected numbers is 23%.[8] This is a nonsignificant result and it indicates that the fit is not so close to expectation that we must reject the hypothesis of race-blind selection from the pool. By this analysis the data are consistent with affirmative action that only gives minority applicants an equal

---

[7] *Id.* at 369.

[8] Technical note: summing 3 df chi-squared across all 6 years yields 22.1 on 18 df. The probability of a chi-squared that small or smaller is $P = 0.23$.

chance with whites. But using another method of calculation, the correspondence is too close to have occurred by chance. Evidently the case for a quota is marginal on these data.

Statistical and Legal Significance

The Supreme Court has said that statistical disparities must be "longstanding and gross" to be sufficient for a prima facie case of discrimination.[9] Differences that are longstanding and gross will also tend to be statistically significant, but not all statistically significant differences are legally significant. It is one thing to have a disparity large enough to reject chance as an explanation and quite another to have one large enough to reject, prima facie, other possible innocent explanations of disparities. The former refers to statistical significance and the latter to practical or legal significance.

The distinction between statistical and practical significance is illustrated by the four-fifths rule of the U.S. Equal Employment Opportunity Commission. As previously stated, in determining whether test procedures used by employers have an adverse impact on minority groups, the Commission uses a rule that a selection rate for any race, sex, or ethnic group that is less than 80% of the rate for the group with the highest rate will generally be regarded as evidence of adverse impact, while a greater than a four-fifths rate will generally not be so regarded.[10] If the numbers are too small for statistical significance, evidence of the test's use over a longer period or by other employers may be considered. By adopting an 80% rule of practical significance and separate instructions on the issue of statistical significance, the Commission has recognized, and indeed emphasized, the difference between the two concepts.

The fact that small differences are not legally significant raises the question whether in testing for significance the null hypothesis ought to be exact equality or merely the largest permissible inequality. For example, since the EEOC permits a ratio of 80% in pass rates before finding that a test has a disparate impact, the significance of a sample disparity might be tested with respect to the null hypothesis that the ratio is 80%. This seems not unreasonable in principle, but meets the practical objection that the sample data would then have to show a disparity greater than 80% in order to find a statistically significant difference as against 80%. To my knowledge no court has adopted that point of view, but it might be appropriate in some contexts. We return to this question in Chapter 6 on confidence intervals.

But some courts, looking for justification in difficult decisions, have conflated legal and statistical significance. Here is an example. The United States brought a civil rights action against the State of Delaware, claiming that the State's use of a literacy examination called Alert discriminated against blacks in the hiring of state

---

[9] Int'l Brotherhood of Teamsters v. United States, 431 U.S. 324, 340, n.20 (1977).

[10] 29 C.F.R. 1607.4 (D) (1998).

troopers.[11] One issue in the case was whether test scores correlated sufficiently with measures of job performance to justify its use, despite its adverse impact on black applicants. An expert for the state compared test scores of incumbent troopers with job performance scores and found correlations between 0.15 and 0.25.

The court then had to decide whether the correlations were high enough to give the test validity. There was "earnest contention" between the experts on whether these correlations could be described as "low" or "moderate." The court rejected "moderate" because a standard statistical text deemed correlations of 0.1 as "low" and 0.3 as "moderate," and none of the statistically significant coefficients had reached the higher level. More usefully, the court pointed out that defendants conceded that when the coefficients were squared, the test scores only explained between 4 and 9% of the variation in trooper performance (which seems too high, given the unsquared coefficients). The court nevertheless found that the test had validity because the coefficients were statistically significant. "Weak though the predictive capacity may be, however, if the strength of the statistical relationship is such that it reaches a benchmark level of statistical significance, then, as...the United States' expert statistician stated, one can conclude that the relationship between the two variables is 'real.'" That is true, but does not answer the legal policy question whether the correlation with performance is sufficient for the court to declare it valid despite its discriminatory impact. As we note below, almost any correlation can become significant if based on a large enough sample.

The court was probably influenced by the fact that Alert was a well established test, used in a number of states. In addition, as the defendants pointed out, the correlations were probably attenuated by the truncation in test scores, which occurred because applicants scoring below 75 on Alert were rejected, never became troopers, and consequently were not included in the data. When adjustment was made for truncation, by a method that was unexplained, but probably questionable, the correlations rose into the low 30s range (e.g., the correlation between test score and composite performance rose to 31%). In its appraisal, the court ignored these higher correlations because their statistical significance could not be calculated. Finally, although it upheld the test, the court found that the passing grade used by the state was too high and directed defendants to reduce it, thereby removing some of the Alert's sting.

Factors That Determine Significance

The size of the sample is the key factor in determining significance. As sample size increases, the standard error of the sample mean decreases by the square root of the sample size. In an employment discrimination model, quadrupling the sample size cuts the standard error of the sample mean wage in half, and this multiplies by two the number standard errors represented by a given disparity. Thus the level

---

[11] United States v. Delaware, 2004 U.S. Dist. Lexis 4560 (D. Del. 2004).

of significance increases as the effective sample size increases and even a small difference will be declared significant if the sample is large enough.

On the other hand, a small sample in which the difference appears to be non-significant is only weak evidence of the truth of the null hypothesis, and so we only say that if the sample is nonsignificant we cannot reject the null hypothesis, not that we accept it. A nonsignificant finding is evidence in favor of the null hypothesis only if there is a large enough sample so that if the null hypothesis were false there would be only a small probability that the sample would be nonsignificant. This is the subject of power, which we discuss in Chapter 7.

I mentioned earlier that other things being equal, significance increased with effective sample size. The effective sample size is the number of *independent* elements in the sample. In the normal employee situation each employee is independent because the salary of one does not determine the salary of another (at least in theory; in fact there may be dependencies). But suppose the issue involves promotions and data from several years are combined. In that case, if the data for an employee twice denied promotion is treated as two employees and two denials, the contribution to effective sample size will not be two because the denials are likely to be correlated. Ignoring the correlation leads to an overstatement of sample size, although the degree of overstatement and its effect on significance are hard to measure.

A third factor that determines significance is the variability of the effect being measured. A blood sample of only a few drops is nevertheless very accurate because blood is a relatively homogeneous fluid; the heart is a good randomization device. Other things being equal, a difference is more likely to be declared significant if the population is relatively more homogeneous than if it is diverse. Thus the difference between male and female executive salaries is more likely to be declared significant than the difference averaged over all employees because salaries within the executive group vary less than over the entire workforce.

Considerations of this sort lead statisticians to sampling in which the population is stratified into more homogeneous subgroups, random samples are taken from each stratum, and results combined. This is discussed in Chapter 8.

A fourth factor in determining significance is whether the statistical test is one-tailed or two-tailed, as we've already discussed. One-tailed tests are more powerful than two-tailed tests – which means that we are more likely to reject the null hypothesis when it is false with a one-tailed test than with a two-tailed test. This suggests that one should prefer one-tailed tests when they are appropriate – as they are in most discrimination cases.

Fifth, we need to consider the interrelation between selection effect and statistical significance. The usual way to adjust for selection effect when there are multiple trials is to use Bonferroni's inequality. This is done in one of two equivalent ways. In one way, the probability of a match is multiplied by the number of trials to get the probability of at least one match; the null hypothesis is rejected only if that probability is less than 0.05. In the other way, the figure 0.05 is divided by the number of trials, and that figure is used as the critical level of significance. In our carpet fiber example from the last chapter, there were 50 trials with a 0.01 probability of matching on each. A single match would not be regarded as significant because an upper bound for the probability of at least one match is 0.50, which is ten times the usual 0.05 critical level. Alternatively, the level of significance needed to reject the null is $0.05/50 = 0.001$, which is smaller than the 0.01 attained significance level of a single match by the same factor of 10.

Adjusting significance levels for multiple comparisons became an issue in a case involving cheating on a police department multiple-choice examination for promotion to sergeant. Anonymous notes handed to the test administrators and other facts suggested that certain named officers had been cheating by illicit communication. An investigation was launched, followed by administrative proceedings against the officers. One pair of identified officers had six and seven wrong answers, respectively, and six of the answers matched (i.e., were the same wrong answers), the maximum possible. A statistician for the Internal Affairs Bureau, which brought the proceedings, testified that in a random sample of 10,000 other pairs of officers who took the test with the same numbers of wrong answers (to any questions, not necessarily to the same questions as the accused pair), none had as many as six matches. He thus computed that an upper bound for the probability of getting six matches by chance as $p = 0.0003$.

The tribunal held that the evidence of the anonymous notes was inadmissible. From this the defense expert argued that the accused officers should be treated as if they had been identified by a search of all possible pairs among the 12,570 officers taking the test – or a monumental 79 million pairs. The expert then applied Bonferroni's inequality to derive an adjusted level of statistical significance reflecting the search. Dividing the level of significance (the expert without explanation used a non-standard figure of 0.001) by that figure gave a cutoff probability of 1 in 79 billion. Since the level of significance was far above that level, the defense claimed that the respondent officers did not have a statistically significant large number of matching wrong answers.

There are two problems with this argument. First, the nature of the evidence that points to the accused officers is irrelevant in determining whether there has been a screening search. They could have been identified by a Ouija board for all that it matters. The only question is whether the statistical evidence was used to make the preliminary identification; if not, there was no screening. But the more fundamental issue is whether an adjustment for multiple comparisons should be made *if* the test-takers had been initially identified by a screening. Such an adjustment would be appropriate if the null hypothesis were whether *any* pairs of officers had a suspiciously high number of matching wrong answers. In that case, there would be a single hypothesis and multiple chances to reject it, so adjustment would be required. But if a suspicious pair is identified by matching wrong answers, the null hypothesis at issue is not whether there are *any* suspicious pairs, but whether the matching wrong answers of *this particular pair* could be attributed to coincidence. Putting aside certain details (to be discussed below) there is only one chance to reject that particularized null hypothesis, and therefore no adjustment for multiple comparisons is warranted.

The defense expert also argued that an adjustment for multiple comparisons was required – a division by eight – because the prosecution expert computed statistical indices for four different groups of defendant officers and used two different tests for each group. In addition, there were morning and afternoon sessions of the test, and matching wrong answers were computed for each session separately. But the defense expert did not think that this required adjustment for multiple comparisons because different data were used for each of the sessions.

Interestingly, all of these points are wrong. The fact that four different groups of officers were defendants creates no need for adjustment because there is not one null hypothesis (that none of them cheated) but separate hypotheses for each group. The fact that two different statistical tests were used would require adjustment only if a rejection by either test would lead to a rejection of the null hypothesis. But the second test was designed to address the possibility that the wrong answers of the accused officers were "distractors," i.e., were more likely to be selected than wrong answers to other questions by other pairs of officers. To that end the second test compared the matching wrong answers for the accused officers with all other pairs who gave wrong answers *to the same questions*. The prosecution expert made it clear that the second test was only a check on the first, i.e., if the null hypothesis was rejected by the first test but not by the second the court might conclude that the questions were distractors and not reject the null. Since the second test would not add to the rejection rate, but only reduce it, there was no increase in Type I error by using two tests and therefore no reason to adjust for multiple comparisons. On the other hand, the separate analyses for the morning and afternoon sessions *could* require adjustment if there were two opportunities to reject the null hypothesis that defendants' matching wrong answers were the product of chance and the session in which the rejection occurred was viewed as irrelevant. The defense expert's point that different data were used for the two sessions was no protection from an increase in Type I error associated with two bites at the apple. As this case illustrates, determining when to adjust for multiple comparisons is not always a simple matter.

Nonsignificant Data

When data are not significant there is a problem melding legal and statistical standards. In a 1993 landmark decision, *Daubert v. Merrill Dow Pharmaceuticals, Inc.*,[12] the U.S. Supreme Court held that under Rule 702 of the Federal Rules of Evidence the trial judge was required to be a "gatekeeper" for scientific evidence. Before allowing such evidence to be heard by a jury, the judge had to determine that the proposed testimony was not only relevant but also "reliable." To be reliable, the personal opinion of a qualified expert was no longer enough; the expert's pronouncements had to be based on scientific knowledge. The problem is that civil cases are to be decided by a preponderance of the evidence, which means that each element of the case must be found by the jury to be more likely than not. More certainty than that is not required and *Daubert* did not purport to change that standard.

---

[12] 509 U.S. 579 (1993).

Data that are not statistically significant would not usually be thought sufficient for a scientific finding, so an expert's opinion based on such data would appear to lack the requisite foundation. But on the other hand, requiring significance at the traditional 5% level would *seem* to be raising the bar beyond a mere preponderance of the evidence, which calls for no more than a 51% probability. We emphasize *seem* because, as previously noted, the two probabilities are quite different: The significance probability is the probability of the evidence given the null hypothesis; the preponderance-of-evidence probability is the probability of the hypothesis given the evidence. But the two are sufficiently related so that requiring significance at the 5% level could well raise the required level of proof above a preponderance.

The argument that requiring significance would improperly raise evidentiary standards in civil cases has resonated with some judges. One such case involved hundreds of civil actions claiming that dietary supplements containing ephedra caused strokes, cardiac injury, or seizure. Plaintiffs' experts relied, among other things, on a study which showed that the odds on a hemorrhagic stroke given ephedra use were more than five times greater than such odds when there was no ephedra use. But because of the small number of ephedra users, the odds ratio was not statistically significant at the usual 5% level (the significance probability one expert computed as 7%). After a *Daubert* hearing on admissibility, the court did not completely preclude the experts from testifying; the judge wrote that to do so would impermissibly raise the required quantum of proof. Instead the court held that the experts could not testify to a causal relation with medical or scientific "certainty" (the usual formulation), but only that ephedra "may be a contributing cause of stroke, cardiac injury, and seizure in some [high-risk] people."[13] While one can sympathize with the court's desire not to preclude suggestive evidence while respecting the Supreme Court's directions, the allowed formulation is so indeterminate as to ephedra's effect and scope that it is hard to see how a jury could rely on it in reaching a rational verdict on class-wide damages for the plaintiffs. Moreover, once unmoored from scientific standards, we have no way to define when scientific testimony can be said to rest on scientific knowledge.

Shall we then ignore significance? The problem is that judges know that samples that are too small are untrustworthy, and they are right. We need some way of drawing a line and intuition is not a reliable way of doing so. The concept of significance should thus be viewed as a standard for measuring the adequacy of data when the sample is small enough to raise questions about its accuracy. But this does not mean that nonsignificant evidence is valueless in law, even though it may be treated as having slight value, if any, in scientific pursuits.

Confronted by such tensions, the courts seem to be moving away from requiring knowledge that would pass scientific scrutiny and retreating to the usual comfortable place: Let the jury hear the testimony with its weaknesses exposed and then render a verdict.

---

[13] *In re* Ephedra Products Liability Lit., 393 F. Supp. 2d 181 (S.D.N.Y. 2005).

Chapter 5
Random Variables and Their Distributions

A *random variable* is the outcome of some random process or experiment and is such that the probability of observing any value or set of values can be known, at least in theory. The set of these probabilities is called the *probability distribution* of the random variable. Suppose that a person is selected at random from a population and his or her height is measured. The height of the person selected is a random variable, and the probabilities of selecting each possible height define the probability distribution of such a variable. Similarly, suppose we pick a random sample of n persons and average their heights. The average height in all possible samples of size n also defines a random variable and the probabilities of each possible average height define the probability distribution of the mean heights in such samples.

Contrary to what the name might suggest, a random variable is not totally wayward and unpredictable. Knowing the probability distribution of a random variable tells us a great deal about it. This knowledge enables us to make useful predictions about random processes, even though randomness prevents us from being completely precise in our descriptions. Returning to the prior example, while we cannot accurately predict the exact height for an individual selected at random from a population, we can say with confidence p that the mean height of a sample of n individuals will lie between specified limits. We can make such statements when the mean and variance of height in the population are known or can be estimated from the sample and because the probability distribution of means of random samples is known in many cases. Similarly, while we cannot accurately predict the number of heads in n tosses of a coin, we can make statements like the probability of at least X heads is p, because the probability distribution of numbers of heads in n tosses of a fair coin is known.

Expectation, Variance, and Correlation

Random variables have theoretical versions of the measures of central location, dispersion, and correlation that we discussed in Chapter 2 for realized data. The definitions are generalizations of those concepts.

M.O. Finkelstein, *Basic Concepts of Probability and Statistics in the Law*,
DOI 10.1007/b105519_5, © Springer Science+Business Media, LLC 2009

The mean value of a random variable is called its *mathematical expectation* or its *expected value*. The expected value of a random variable is a weighted average of all possible values of the random variable, with the weights given by the probabilities of the values. Suppose we have a coin to which we assign the value 1 for heads and 0 for tails and which has a probability p of landing heads and $q = 1 - p$ of landing tails. Then the expected value of a single toss is $1 \times p + 0 \times q = p$.

The expected value of the sum of two random variables is the sum of their expectations, irrespective of the correlations between them. So if we toss a coin n times the expected value of the number of heads is n times the expected value of a single toss, or np. The expected value of the proportion of heads in n tosses is $np/n = p$.

The definitions of variance and standard deviation of a random variable are similar to the definitions for data, again with mathematical expectation substituted for averages. Hence the variance is the expected squared deviation of the variable from its expected value and the standard deviation is the positive root expected squared deviation. Applying this definition, the variance of heads on a single toss is $(1-p)^2 p + (0-p)^2(q) = pq$. The variance of the sum of independent variables is equal to the sum of their variances. So if we toss a coin n times, the variance of the number of heads is n times the variance of heads on a single toss, or npq; the standard deviation is $\sqrt{npq}$. The variance of the proportion of heads is easily shown to be pq/n and the standard deviation is the square root of that. These important results are used in the discussion of the binomial distribution below.

Chebyshev's inequality holds for random variables as it does for data. Thus the probability that *any* random variable would depart more than k standard deviations from expectation is less than $1/k^2$. As we previously noted, much stronger statements can be made when the probability distribution of the random variable is known.

The correlation coefficient for two random variables is defined in the same way as for sample data, with mathematical expectation replacing averages in the definitions of covariance and standard deviations of the variables. Certain properties have already been noted.

In most situations in which probabilities are calculated, the probability distributions of random variables are given by formulas, which assume an idealized situation. Some of the most widely used distributions are described below.

The Binomial Distribution

Assume a sequence of n random trials, with n being fixed in advance. Each trial may have one of two outcomes, with the probability of a "positive" outcome remaining fixed throughout the trials and being independent of prior outcomes. Trials in such a sequence are called "Bernoulli trials." Coin tosses are a classic example of Bernoulli trials. Each toss is a trial with a binary outcome (heads or tails), the probability of a head does not change over a run of trials, and the trials are independent. As we have seen, if there are n trials with probability p of success on a single trial, the expected number of successes is np and the variance of the number of successes is

$np(1 - p)$. Although the probability of any particular number of successes is small, the probability is greatest for the expected number and diminishes as the numbers move away from expectation. If a fair coin is tossed 100 times, the expected number of heads is $100 \times 0.50 = 50$. The probability of that result is about 0.08. By contrast, the probability of 0 or 100 heads is fantastically small – less than 1 in 10^{30}. Binomial probabilities have been intuited, computed, or approximated in many discrimination cases.

Avery v. Georgia,[1] is an example of the old style intuitive assessment. In that case a black man was convicted by a jury selected from a panel of 60 veniremen. The panel members were drawn, supposedly at random, from a box containing tickets with the names of persons on the large jury roll – yellow tickets for blacks and white tickets for whites. Five percent of the tickets were yellow, but no yellow tickets were included in the venire of 60. Writing for the U.S. Supreme Court, Justice Frankfurter intuited that "[t]he mind of justice not merely its eyes, would have to be blind to attribute such an occasion to mere fortuity." The conclusion is right, but the probability of zero yellow tickets is not outlandishly small. The selection of veniremen can be modeled, approximately, as a series of 60 Bernoulli trials with a 5% probability of selecting a yellow ticket on each selection. The expected number of yellow tickets is $0.05 \times 60 = 3$. The probability of selecting 60 white tickets at random from a large jury roll that has 5% yellow tickets is less than $0.95^{60} = 0.04$, or 4%. This is just small enough to cause us to reject the null hypothesis of random selection.

Of particular interest in applied work are the "tails" of the binomial distribution. The left tail is the probability of seeing so few "successes" and the right tail is the probability of seeing so many. The "so few" and "so many" event probabilities are *cumulative binomial probabilities*, which have a *cumulative binomial distribution*. The cumulative binomial probability of X or fewer successes is equal to the sum of the probabilities of 0, 1, 2,..., X successes. If the probability of such an extreme result is low, the hypothesis of a binomial distribution is rejected.

In another early jury case, *Whitus v. Georgia*,[2] a venire of 90 was selected "at random" in two stages from a tax digest that was 27% black. The digest indicated race. There were only 7 blacks on the venire when about 24 would have been mathematically expected. The cumulative binomial probability that so few blacks would be selected is equal to the sum of the binomial probabilities of selecting 0, 1, 2,...,7 blacks when there were 90 selections and the probability of selecting a black was 27% each time. The Supreme Court computed this probability as the minuscule 0.0000045, and for this and other reasons rejected the notion that the jury selection was random with respect to race.

The binomial distribution assumes that the probability of success on a single trial remains the same in a series of trials. But if one picks from a finite population, unless each selection is replaced with a person with the same relevant characteristic,

---

[1] 345 U.S. 559 (1953).
[2] 385 U.S. 545 (1967).

the probabilities of success will change as the population changes. Taking account of this leads to the hypergeometric distribution, which is discussed in the next section.

More recent applications of the binomial distribution to jury selection have involved the Sixth Amendment requirement that federal jury panels be drawn from a source representing a "fair cross-section" of the community in which the defendant is tried. In many district courts, the master wheel from which jurors are drawn tends to underrepresent blacks and Hispanics because it is chosen primarily, if not exclusively, from voter registration lists and those groups tend to underregister in proportion to their numbers in the population. To test whether the underrepresentation is sufficient to sustain a claim of violation of the fair-cross-section requirement, courts have looked at the numbers of blacks and Hispanics that would have to be added to the average venire to make it fully representative. This has been called the absolute numbers test. The trouble with this test, as the courts have recognized, is that if the minority populations are small, the numbers needed to be added to a venire will be correspondingly small and so an underrepresentation of small minorities might never be regarded as of consequence.

An alternative approach is to compute the probability that the number of minorities on a venire would not exceed some small number, given the existing wheel, and compare that with the same probability given a fully representative wheel. If the difference in probabilities seems significant, the underrepresentation of minorities in the actual wheel is deemed to deny a fair cross-section. In *United States v. Jackman*,[3] the unrepresentative wheel had 3.8% blacks when there should have been 6.3%. Only 2.5 blacks would have had to be added to an average venire of 100 to make it fully representative. On the other hand, using the binomial distribution, the probability that there would be no more than a single black on a venire was about 10.3% for the unrepresentative wheel, but only 1.1% for a representative wheel. Based on calculations of this kind, the court found that the unrepresentative wheel was not a fair cross-section.

Because jurors vote in one of two ways in criminal trials, guilty or not guilty, the process has a very superficial resemblance to a binomial model. However, the probability of a guilty vote is not constant, but varies from case to case and from juror to juror in a single case, and the votes certainly are not independent. These are crucial distinctions, especially the latter. Nevertheless, mathematicians, beginning with Nicolaus Bernoulli in the eighteenth century, have used binomial models to model jury behavior, usually uncritically ignoring these complications. Perhaps the nadir of this kind of effort was Justice Blackmun's opinion for the Court in *Ballew v. Georgia*.[4] In *Ballew*, the question was whether five-person juries in criminal trials, provided for by a Georgia statute, violated the Sixth and Fourteenth Amendments. Although in a prior case the Court held that six-person juries were constitutional, finding that there was "no discernible difference" between the results reached by different-sized juries, in *Ballew* the Court held that size mattered. In reaching that

---

[3] 46 F.3d 1240 (2d Cir. 1995).
[4] 435 U.S. 223 (1978).

conclusion, the Court relied, in part, on an article by two social scientists who had purported to calculate error rates in juror verdicts for different-sized juries, based on a binomial model. In their article, the authors made, as they put it, "the temporary assumption, for the sake of calculation," that each juror voted independently (the coin-flipping model, as they called it) and the assumptions "for sake of discussion," that 40% of innocent defendants and 70% of guilty defendants would be convicted. From these assumptions, using the binomial distribution, they calculated that the probability that a juror would vote (i) to convict an innocent person was 0.926 and to convict a guilty person was 0.971 and (ii) to acquit a guilty person was 0.029 and to acquit an innocent person 0.074. They then used these probabilities in a binomial model, adding the assumption that 95% of defendants were guilty and the assumption that, following Blackstone, the error of convicting an innocent person was ten times worse than acquitting a guilty one. This enabled them to come up with verdict error rates for different-sized juries. The point of minimum weighted errors was between six and eight persons, and "as the size diminished to five and below the weighted sum of errors increased because of the enlarging risk of convicting innocent defendants." But even crediting the model, the difference it showed was negligible: 468 errors per 1,000 at 7 vs. 470 errors per 1,000 at 5. Nevertheless, the Court cited the study in support of its conclusion that five-person juries would be less accurate than larger juries.

Given the arbitrariness of the assumptions, the patent inappropriateness of the binomial model, and the outlandishness of the computed probabilities of individual juror and jury errors, it is remarkable, and unfortunate, that the Court ignored the caveats of the social scientists and embraced the results of their calculations as expressing truths about the effect of jury size on the correctness of verdicts.

The Hypergeometric Distribution

As previously stated, the binomial distribution applies to a series of trials with a constant probability p of success on each trial. This is no longer true when, as in jury selection, the trials consist of selections made from a finite population and the items selected are not returned to the population before the next selection. In that case the probability distribution of the number of successes is no longer binomial, but *hypergeometric*. Such probabilities are used in the same way as binomial probabilities to test hypotheses; in particular, cumulative hypergeometric probabilities are analogous to cumulative binomial probabilities.

Hypergeometric and binomial distributions have the same expected value, np, where n is the number of trials and p is the probability of success at the outset of the trials. However, the variance of a hypergeometric distribution is smaller than npq, the variance of the binomial distribution, because over a run of trials disproportionate results in one direction raise the probability of subsequent results in the other direction, so that the outcomes tend to vary less from expectation than in the binomial case. The adjustment is made by multiplying the binomial variance by a *finite population correction factor*, given by $(N-n)/(N-1)$, where N is the size of the population and n is the size of the sample selected from it. Note that if a single selection is made from the population, $n = 1$, the finite population correction factor is 1, so there is no reduction in the variance, as one would expect. As the size of the sample approaches the entire population the correction factor approaches 0; there is no more room for variability. The difference between the binomial and hypergeometric variances is not great, and the binomial is frequently used as an approximation for the hypergeometric when the sample is small relative to the population. When it is large, however, the difference can become important. The formula shows us that the reduction in variance is about in the proportion that the number selected bears to the total population; if 10% of the population is selected, the variance of the number of successes is reduced by about 10%.

There is another justification for using the binomial distribution in the jury discrimination cases despite the fact that in most cases the distribution of black jurors is really hypergeometric. Most calculations of probabilities are made by defendants who want to show the small probabilities of so few minorities under the hypothesis of random selection. In that context a defendant's use of the binomial is conservative because the probabilities would be even smaller if the hypergeometric distribution had been used instead. In that context, a prosecutor's use of the binomial would not be conservative.

Miller-El v. Cockrell[5] is a case decided a few years ago by the Supreme Court. A jury in Dallas Country, Texas, convicted Thomas Joe Miller-El of capital murder and the trial court sentenced him to death. At his trial the prosecutors used peremptory strikes to exclude 10 of the 11 blacks eligible to serve on the jury, but only 4 of the 31 prospective white jurors. After many years of appeals, the Supreme Court held (8–1) that these statistics created a colorable claim of race-based use of peremptory challenges, noting that "happenstance is unlikely to produce this disparity." That was an understatement. The null hypothesis of random selection with respect to race can be modeled by assuming that there were 11 black chips and 31 white chips in an urn and that 14 of them were selected for peremptory challenge at random from the urn. The expected number of black chips selected for a strike is about 3.7. The cumulative hypergeometric probability that 10 or 11 of the selected chips would be black is about 1 in 151,000; the null hypothesis must be rejected. If the binomial distribution is used instead of the hypergeometric, the probability of so many black chips is much larger, about 1 in 2,000. It is, of course, still very small. Miller-El could have used either distribution to make his point.

Statistical analyses in disputed election cases also provide examples of the hypergeometric distribution. Under New York law a defeated candidate is entitled to a rerun if improperly cast votes (it being unknown for whom they were cast) "are sufficiently large in number to establish *the probability* that the result would be changed by a shift in, or invalidation of, the questioned votes." In determining this probability, one can again model the process by assuming that all the votes cast in

---

[5]537 U.S. 322 (2003).

the election are chips in an urn, marked with the names of the candidates for whom they were cast. Chips are withdrawn at random from the urn, an operation that corresponds to their invalidation. The number of the winner's chips is a random variable that has a hypergeometric distribution because chips are not replaced after being selected. The probability that a sufficiently large number of chips for the winning candidate would be withdrawn so that the number of chips for the winner remaining in the urn would no longer exceed the other candidate's chips is the cumulative hypergeometric probability that the election would be changed by the invalidation.

In one election case, *Ippolito v. Power*,[6] the ostensible winner had a plurality of 17 votes; among the 2,827 votes cast there were 101 invalid votes. If at least 59 out of the 101 invalid votes had been cast for the winner, the removal of invalid votes would have changed the result of the election. This seems like something that could easily happen since the expected number of random invalid votes for the winner was 50.8. But if the invalid votes were indeed randomly distributed among all votes, the cumulative hypergeometric probability of that result is only about 6%. Because the sample (101) was small relative to the population (2,827) there was very little difference between the hypergeometric and binomial probabilities. The winner, who wanted to show a small probability of reversal, could have used a binomial calculation because it would have shown a slightly larger probability of reversal. The value of probability methods in this context is to correct the misleading intuition that invalid random votes somewhat exceeding the winner's margin create a significant probability of reversal, and to direct attention to the important issue in most such cases, namely, whether the invalid votes were likely to be skewed for one candidate or the other, as opposed to a random sample.

The Normal Distribution

The normal distribution is the familiar symmetrical bell-shaped curve that is a key tool in mathematical statistics. Unlike the binomial distribution, which applies only to integers and is most accurately represented by a histogram with vertical bars for the numbers of successes, the normal distribution is a continuous distribution that can be represented by a curved line. See Fig. 5.1. There is a central hump (the mean of the distribution), with symmetrical tails that approach zero as they go to infinity on either side. The height of the curve at any given point compared with the height at another point represents the ratio of the relative frequency of the x-axis value at the given point compared to that at the other point. For that reason, the figure is sometimes referred to as a relative-frequency curve. The area under the curve between any two x-values is the probability that a random variable with a normal distribution would lie in that range.

There is in fact a family of normal distributions that differ in their mean and standard deviation. The common thread among them is that a random variable with a

---

[6]22 N.Y. 2d 594 (1968).

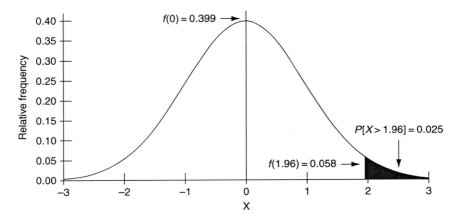

Fig. 5.1   The standard normal distribution

normal distribution (whatever its mean and standard deviation) has the same probability of departing from the mean by X or more standard deviations. About 68% of the area under the curve lies within one standard deviation of the mean; about 90% lies within 1.64 standard deviations, and about 95% lies within 1.96 standard deviations. Figure 5.1 shows that about 2.5% of the area under the curve lies more than two standard deviations above the mean and another 2.5% more than two standard deviations below the mean. The often-seen quotation of a mean value plus or minus two standard deviations is based on the fact that a normal variable falls within two standard deviations of its expected value with a probability close to the conventional 95%.

The curve was originally known as the astronomical error law because it described the distribution of measurement errors in astronomical observations. The fact that events as wayward and unpredictable as errors have a knowable probability distribution caused a sensation when it was discovered and explored by Carl Fredrich Gauss (1777–1854) in the early nineteenth century. The normal distribution is frequently called Gaussian in honor of his work. Subsequent investigators extended the concept of a law of errors from the narrow context of measurement error to other areas of natural variation. Adolphe Quetelet (1796–1874), a Belgian astronomer and poet, claimed that variation in human characteristics was normally distributed because they were caused by accidental factors operating on a true or ideal type – *l'homme type* – the average man. Quetelet became famous for this concept. Initially he viewed the idea merely as a device for smoothing away random variation and uncovering mathematical laws of social behavior. But later, influenced by his conclusion that variation represented accidental imperfections, Quetelet endowed the average man with superior moral qualities emblematic of democratic virtue. The tendency to overestimate the prevalence of the normal distribution has been dubbed "Quetelismus."

The most often used member of the normal family is the *standard normal distribution*, which has a mean zero and a unit standard deviation. A table of the standard normal distribution is printed in most statistics books and is available on many hand-held calculators. The table usually permits a calculation of the tail probabilities associated with departures of between 0 and 4 or 5 standard deviations from the mean. To compute a tail probability associated with a departure from the expected value under some hypothesis, the usual practice is to compute the departure in terms of the number of standard deviations and then to refer to the standard normal table.

How is it possible that something as seemingly unpredictable as errors have a definite mathematical form? The answer comes from various remarkable *central limit* theorems. In substance, it can be shown that the sum of many independent factors, each of which has only a slight effect on the outcome of the process as a whole, will approach a standard normal distribution (when suitably adjusted), as the number of factors becomes large. The adjustment usually made is to subtract from the sum its expected value and to divide by its standard deviation. The result is a standardized random variable with zero mean and unit variance, whose distribution approaches the standard normal distribution as the number of components becomes large. The central limit theorem tells us that measurement errors will have a normal distribution if they are the sum of many small influences. It is the central role played by the normal distribution, as the limiting distribution of standardized sums of many other distributions, that gives this particular limit theorem its name and its importance.

A prime example is the sample mean. Since it is the sum of a number of independent components making up the sample, in sufficiently large samples the sample mean will have an approximately normal distribution about its expected value. An equivalent formulation is that the number of standard errors between the sample mean and its expected value follows a standard normal distribution, in the limit as the sample size becomes large. For samples that are large enough, this result holds whether or not the underlying distributions of the components are themselves normal. In particular, a binomial or hypergeometric variable will be normally distributed if the number of trials is sufficiently large for the central limit theorem to work. For example, returning to *Whitus v. Georgia*, the expected number of blacks on the venire of 90 was, as stated, about 24. The actual number was 7, the difference being 17. The standard deviation of the number of blacks was $\sqrt{90 \cdot 0.27 \cdot 0.73} = 4.2$. The departure from expectation is $17/4.2 = 4.1$ standard deviations. A table of the normal distribution tells us that the probability of being at least 4.1 standard deviations below expectation is 0.00002. This is larger than the exact computation, but is sufficiently close to it for practical purposes.

In the past, statisticians and others tended to assume, reflexively, that data were normally distributed and made calculations on that basis. Today professional statisticians and economists are more sensitive to possible departures from normality, but one sees the less discerning attitude in non-statisticians who, using statistical tools, continue to make the old assumptions.

One tax case involved the question whether certain configurations of long and short futures contracts involving silver, called silver butterfly straddles, held for 6 months, moved enough in price during that period to permit profitable liquidation after commissions. The IRS took the position that since butterfly straddles were very unlikely to be profitable, they were not legitimate investments, but merely tax dodges, and so the deductions they generated should be denied. At trial, an IRS expert (not a professional statistician) testified that because commissions associated with buying and liquidating butterfly straddles exceeded two standard deviations in price, and since prices of straddles held for 6 months were normally distributed, it was improbable that prices would fluctuate sufficiently to cover such costs and allow a profit even if they moved in a favorable direction. The assumption of normality was justified because the price movement over the 6-month period represented the sum of one-day movements; by the central limit theorem this sum should be normally distributed. However, an expert for the taxpayer tested the price data and found that in fact they were not consistent with a normal distribution, but had "heavy" tails, meaning that price movements more than two standard deviations from the mean were more frequent than would be expected under a normal distribution. This occurred for two reasons: day-to-day price movements were not independent and some days contributed very large movements, contrary to the assumption of many small component influences. This case illustrates an important point: It is not uncommon for there to be more variability in data than would be predicted by the normal distribution.

The Poisson Distribution

Consider X, the number of events occurring in n independent trials with constant probability of success p, so that X has a binomial distribution with mean np. Let n become large while p approaches 0 in such a way that np approaches a constant μ. This might occur if we consider the probability of events – such as accidents – occurring over some time period. As we divide the time interval into progressively smaller segments, the probability p of an event within each segment decreases, but since the number of segments increases, the product may remain constant. In such cases, X has a limiting distribution known as the Poisson distribution after the French mathematician Simeon Denis Poisson (1791–1840). Because it describes the distribution of counts of individually rare events that occur in a large number of trials, many natural phenomena are observed to follow this distribution. The number of traffic accidents at a busy intersection and the number of adverse events in a vaccination program are examples. The Poisson distribution gives the probabilities of various numbers of events, given a mean number μ. In a Poisson process the likelihood of future events does not depend on the past pattern of events.

For example, in one case plaintiffs sued the Israeli government when their child suffered functional damage after receiving a vaccination in a government program. The vaccine was known to produce functional damage as an extremely rare side effect. From prior studies in other programs it was known that this type of vaccine had an average rate of 1 case in 310,000 vaccinations. Plaintiffs were informed of this risk and accepted it. The child's vaccine came from a batch that caused 4 cases out of about 310,000. Was this consistent with the risk that plaintiffs accepted? The answer is given by the Poisson distribution from which it can be computed that, given an expected number of 1 case per 310,000 vaccinations, the probability of 4 or more cases is about 1.7%. The hypothesis that the batch of vaccine used in this program was no riskier than prior batches would have to be questioned.

Student's t-Distribution

Thus far we have been comparing proportions. Many problems, however, involve a sample mean rather than a sample proportion. The central limit theorem gives us some comfort that, if the sample is large enough, the sample mean, like the sample proportion, suitably adjusted, will have a standard normal distribution. The adjustment is to subtract the expected value under the null hypothesis and divide by the standard error of the sample mean. In the case of proportions, the null hypothesis, which specifies a p-value for success on a single trial, also specifies the standard error of the sample proportion. But in the case of the mean this is not so: A mean specified by the null hypothesis does not specify the standard error, which could be anything. Instead, in most cases, we must estimate the standard error from the sample itself. In relatively large samples there is no difference; the sample mean is still normally distributed. But if the sample is small to moderate (the figure usually given is less than 30), the need to estimate the standard error itself introduces additional, and non-negligible, sampling variability, which makes the tails of the distribution fatter than those of the normal distribution. One W.S. Gosset, who wrote under the name Student, intuited the mathematical form of the sampling error of the mean when the standard deviation of the population was estimated from the sample drawn from a normal population. This distribution is called *Student's t* in his honor. The *t-value* of a statistic is its value divided by its standard error, with the latter estimated from the data. The thickness of the tails depends on the size of the sample: The smaller the sample, the greater the sampling variability compared with the normal distribution. There is thus a family of *t*-distributions, depending on what are called degrees of freedom, which equals the sample size less one. Tables of the *t*-distribution appear in statistics textbooks. A test of a null hypothesis based on the *t*-table is unsurprisingly called a *t-test*. For example, in a sample of size four, the degrees of freedom are three. The probability of a *t*-value of 1.96 or more using the *t*-distribution is 6.5%, compared with 5% under the normal distribution. Student's *t*-distribution for three degrees of freedom compared with the standard normal distribution is shown in Fig. 5.2.

Here is an example. The Federal Clean Air Act requires that before a new fuel or fuel additive may be sold in the United States, the producer must demonstrate that the emission products will not cause a vehicle to fail certified emission standards. To estimate the difference in emission levels, a sample of cars is driven first with the standard fuel and then with the new fuel and the emission levels compared. In testing a new fuel called Petrocoal, a gasoline with a methanol additive, 16 cars were

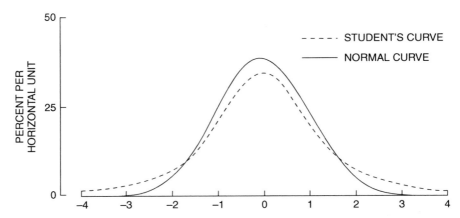

Fig. 5.2 Student's *t*-curve. The *dashed line* is Student's curve for three degrees of freedom. The *solid line* is a normal curve, for comparison

driven first with a standard fuel and then with Petrocoal, and the nitrous oxide levels compared. The data showed a range of differences for the 16 cars when Petrocoal was used, from an increase of 0.343 to a decrease of 0.400. There was an average increase with Petrocoal of 0.0841, with an estimated standard error of 0.0418; thus the average increase in nitrous oxide when Petrocoal was used was 0.0841/0.0418 = 2.01 standard errors above expectation (zero increase). This would be significant by the usual standards using the normal distribution, but since we estimated the standard error of the increase from the data, we must use the *t*-distribution. The *t*-table tells us that, for 15 degrees of freedom, the probability of a *t*-value of at least 2.01 is only about 6.6%. But because the EPA used a 10% level of significance, it found that the excess emissions were significant and rejected Petrocoal.

The Geometric and Exponential Distributions

Many problems in the law relate to how long someone, or something, will live (or last). When there is a series of discrete independent trials, with probability p of failure on each trial, the probability distribution of the number of trials to the first failure is described by what is called the *geometric distribution*. In trials with a geometric distribution, the mean number of such trials is given by $1/p$. Thus, if there is a 5% chance of failure on a single trial, the mean number of trials to failure is 20. Unlike the distributions discussed thus far, there is no "bunching" of probability around the mean number. In fact, the single result with the greatest probability of failure is not the mean number of trials but the first trial. See Fig. 5.3.

When the risk of termination is continuous, as in the life of a wine glass in a restaurant, or an account that may be terminated at any time, the analogue to the geometrical distribution is the *exponential distribution*. In such cases, the analogue

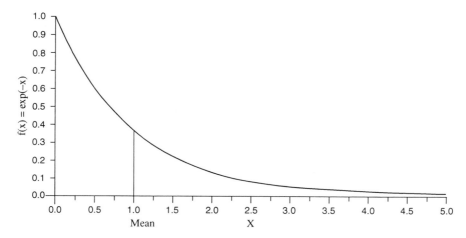

Fig. 5.3 Probability density function for the standard exponential distribution

of p is the hazard constant, β, which is defined as the limit of the conditional probability of failure in a given brief time interval, divided by the length of that interval, as the length shrinks to zero. The hazard is called "the force of mortality" and is like the instantaneous velocity of a car. The hallmark of the exponential distribution is that the hazard remains constant over time. The mean waiting time is $1/\beta$ and the standard deviation is also $1/\beta$. As Figure 5.3 shows, this probability distribution is not symmetrical around the mean. The mean waiting time is only one standard deviation from the left end of the probability density curve (0 waiting time) while the right tail extends to infinity. The heavier right tail pulls the mean waiting time to the right of the median; specifically, the median is only about 69% of the mean. This tells us that the probability of failure at times below the mean is greater than at times above the mean.

The geometric and exponential distributions are special cases of more general models that allow for a hazard that is not constant, but rather changes over time. This, of course, is common in applied work. The so-called *Weibull* model is a prime example. In one case a company sold paint for house sidings with a 5-year warranty. When the paint started to fail, warranty claims were made against the seller, who in turn sued the manufacturer. In the seller's suit, the court required proof of damages arising from all past and future failures, not only those arising up to the time of trial. The data on all failures prior to trial indicated that the risk of failure grew with time. A Weibull model was fitted to the data on houses for which the paint had already failed and used to project the probability that the paint would last 5 years (which was very small). Damages were computed accordingly.

Using the right model can make a big difference in acceptance of mean life estimates. Two similar cases with contrasting outcomes illustrate this.

In a tax case, the purchaser of a television station sought to depreciate for tax purposes the part of the purchase price (millions of dollars) allocated to the station's network affiliation contract. (An affiliation contract between a television station and a television network permits the station to broadcast network programs; in days before cable and satellite a contract with a leading network could be the station's most valuable asset.) Under the tax law, an intangible asset, such as a contract, could be depreciated only if its "life" could be estimated with "reasonable accuracy." The contracts were 1- or 2-year agreements; they were usually, but not always, renewed. The purchaser attempted to estimate a life for its contract by looking at industry statistics of contract lives. A median life could be estimated because more than half the contracts had already terminated, but a mean life could not be directly estimated because many contracts had not yet terminated. The purchaser had to use the mean life to make its argument that allowing the deduction would involve no loss to the Treasury, since the mean could be substituted for each actual life and the total years of life would be the same. From industry statistics, the taxpayer calculated that the risk of nonrenewal was about 5% at each renewal time and using a geometric distribution model calculated a 20-year life for its contract. But the court rejected the assumption of a constant risk of termination, finding instead that the risk of termination declined with successive renewals, as the relationship solidified. Since the taxpayer had failed to estimate the life of the contract with reasonable accuracy, the deduction for depreciation was denied.[7] This was an unusual situation since the risk of failure usually increases rather than decreases with age.

The same problem arose when a bank that acquired another bank wanted to amortize for tax purposes the portion of its acquisition costs allocated to the savings accounts of the acquired bank. The IRS denied any deduction on the ground that the life of the accounts could not be estimated with reasonable accuracy. The bank went to court and at the trial an expert for the bank testified to a Weibull model, estimated from the data, which showed a declining hazard over time. From the model he computed a mean life for the accounts and the bank won the case.[8]

---

[7] Commissioner of Internal Revenue v. Indiana Broadcasting Corp., 350 F.2d 580 (7th Cir. 1965).
[8] Trustmark Corp. v. Commissioner of Internal Revenue, T.C. Memo 1994-184 (1994).

Chapter 6
Confidence

When William Hill was arrested possessing 55 plastic bags containing white powder, the severity of Hill's offense depended on the amount of cocaine he possessed. Government chemists picked three bags at random, tested them, and found that they all contained cocaine. If the bags had been pills in a bottle it has been held permissible to presume, based on a random sample, that they all contained cocaine.[1] But at least some courts do not allow that presumption for separate bags.[2] Assuming that is the situation in Hill's case, what inference can we draw as to the number of bags among the 55 that contained cocaine? From the random sample of three, we want to estimate the contents of the population of bags in Hill's possession. This is the problem of statistical *estimation* and it leads to the concept of confidence in the following way.

Of course at least 3 of the 55 bags contained cocaine (bags with cocaine are referred to as "c-bags"). But if there were *only* three c-bags the probability that they all would have been selected in three random drawings is vanishingly small. It is in fact less than 1 in 26,000. So three is an unreasonably low number for our estimate. What is the smallest number of c-bags that would not be ruled out, given that three random selections were all c-bags? The question can be put in terms of testing a null hypothesis: What is the smallest number of c-bags among the 55 that we would not reject as a null hypothesis given three random selections that were all c-bags? Applying the usual 5% criterion for testing, a calculation shows that if there were 21 c-bags there would be a 5% probability of picking 3 c-bags at random. If there were fewer than that, the probability of picking 3 of them is less than 5%, and so we would reject any lower number as a null hypothesis. The number 21 is a lower bound estimate for the population number of c-bags. Our *95% one-sided confidence interval* is thus from 21 to 55, which consists of all those hypotheses that cannot be rejected at the usual 5% level, given our sample results. Or, to use other words, we are at least 95% *confident* that there are no fewer than 21 c-bags, since there is less than a 5% probability of such a sample result if there were in fact fewer than 21.

---

[1] People v. Kaludis, 497 N.E. 2d 360 (1996).

[2] People v. Hill, 524 N.E.2d 604 (1998).

M.O. Finkelstein, *Basic Concepts of Probability and Statistics in the Law*, DOI 10.1007/b105519_6, © Springer Science+Business Media, LLC 2009

Of course 95% is simply an arbitrary conventional level of significance. We could have picked a higher (e.g., 99%) or lower (e.g., 90%) level of confidence. The lower bound of the interval would have to be lower to allow a higher level of confidence, or could be higher if a lower level of confidence is acceptable. If we had insisted on a 99% confidence interval, the lower bound would be 12; at the 90% confidence level the lower bound would be 26.

Here is another example that, conceptually, is very similar. A company named Financial Information, Inc., collected and published newspaper notices of the redemption of municipal and corporate bonds and distributed them to its subscribers. Another, much larger company, Moody's Investor Services, published similar information. Financial discovered that in its notices with typographical errors, the errors also appeared in Moody's notices. The congruence was remarkable: In a 2-year period 15 out of 18 notices with errors were reproduced by Moody's. This led Financial to sue Moody's, claiming wholesale appropriation of its work product. Moody's did not dispute that it had copied those Financial notices with errors, but its officers testified that it copied only about 22 notices a year out of some 600 in which the information was first published by Financial and then by Moody's. Is that testimony believable?

A statistician appearing for Financial testified in substance as follows: Assuming that the errors occurred at random, the 18 error notices were a random sample from the 600 for which we can determine copied status. To compute a 99% one-sided confidence interval, we calculate the smallest proportion of copied notices in the 600 for which the probability of selecting 15 or more copied notices out of 18 in a random sample is not less than 1%. The answer is 0.54, or over 300 notices. The position that only 22 notices were copied is not tenable. As in the cocaine problem, the statistician computed a lower bound for a one-sided confidence interval, this time a 99% interval.

Despite the statistical evidence, which was unrebutted, the district court found for Moody's because it believed the Moody's executives. The court of appeals reversed because the district court failed to consider the testimony of Financial's statistician cited above. On remand, the district court again found for Moody's and on the second appeal was affirmed by the court of appeals. The ground was that the notices were not copyrightable and there was preemption of state unfair competition laws. In passing, the appeals court noted that the record would support a finding of *no* wholesale appropriation by Moody's because (i) Financial's expert could be "statistically certain that Moody's had copied only 40–50% of the time" and (ii) Moody's showed that of its 1,400 notices in one year, 789 could not have been copied by Moody's.[3] What would have been wholesale appropriation was not defined.

Why speak of confidence rather than probability? Why don't we say that there is a 95% probability that the number of c-bags is at least 21 or that there is a 99% probability that the proportion of copied notices is at least 54%? That is what lawyers

---

[3] Financial Information, Inc. v. Moody's Investors, 751 F.2d 501 (2d Cir. 1984), *aff'd after remand*, 808 F.2d 204 (2d Cir. 1986).

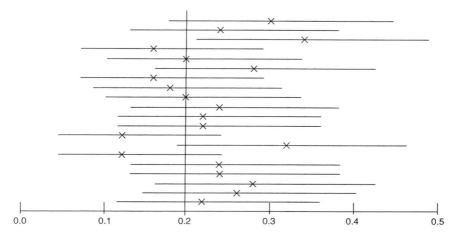

Fig. 6.1 Two-sided 95% confidence limits for p in binomial samples of size 50 when $p = 0.2$

would like to hear. The answer is that what we have calculated by our procedure is the probability that our confidence interval will include the "true" or population value, not the probability that the true value is included in a confidence interval. Figure 6.1 shows this approach: repeated trials with 50 flips of a weighted coin ($p = 0.2$ for success) in each trial, the proportions of successes marked by $\times$'s, with horizontal lines for the 95% two-sided confidence intervals. In 19 out of 20 of the trials the lines cover the true value of 0.2. The inverse calculation (the probability that the true value lies within the confidence interval) involves the Bayesian assumption that the confidence interval is fixed or given and it is the true value that has a probability distribution. Bayesian methods are seldom used in this context because there is no adequate basis for selecting a probability distribution for the true value.

It follows that the classical calculation does not permit us to say that the true value is more likely to lie at or near the sample value than at the edges of the confidence interval, or that it as likely to lie above the sample value as below it. It *is* permissible to say that the sample value is in general a better estimator of the population value than the edges of the confidence interval because the sample result is more likely to be closer to the population value than the edges would be. In addition, if the confidence interval is symmetrical, the sample value is as likely to be above the true value as below it (but this no longer holds if the intervals are asymmetrical, as in Fig. 6.1). The point is that, in the classical calculation, probabilistic statements can be made about the interval, not the true value. Thus confidence intervals cannot be directly translated into statements about the probability of assertions that lawyers (and others) would like to make. However, here is some solace: As the sample size increases, the sample result closes in on the mean or true value and the classical and Bayesian approaches become equivalent.

What we have computed, and what is of interest in the foregoing cases, is the lower bound of a one-sided confidence interval. Using a lower bound, as opposed to the sample result, is appropriate because it is conservative against the party offering the evidence. If the point estimate had been used in the cocaine case, all 55 bags would be deemed to have cocaine compared with the lower bound estimate of 21 bags. In the Financial case, the point estimate is 15/18 = 83% copied notices compared with the lower bound estimate of 54%. So the lower bound estimate is quite conservative in these cases because the samples are small.

One-sided confidence intervals are justified when only one side is of interest because the claim only involves one side (the smallest number of c-bags or the smallest proportion of copied notices). In that case the rate of errors is measured by a one-tailed test, which means a one-sided confidence interval. Two-sided confidence intervals with both lower and upper bounds are common in scientific pursuits and are appropriate in the law when the width of the interval is used to measure the reliability of the sample estimate and a departure in either direction would cast doubt on its adequacy. (We discuss this latter use below.) A two-sided, 95% confidence interval consists of all those points below and above the sample estimate that would not be rejected as a null hypothesis by, in each case, a 2.5% (one-tailed) test. Constructed in this way, 95% of the intervals would cover the true value. One side of two-sided confidence interval will always be wider than a one-sided interval, other things being equal, because in constructing a two-sided interval the test for the null hypothesis is at the 2.5% level (one-tailed) for each bound, whereas in fixing a one-sided (lower or upper) bound the null hypothesis is tested at the 5% level (one-tailed). A 2.5% test does not reject some null hypotheses at the margins that would be rejected by a 5% test. Because one-sided confidence intervals will be narrower, they should be used whenever that is appropriate.

Given the definition of a confidence interval, it should be clear that confidence intervals can be used as tests of significance. A point that is outside a 95% confidence interval is a null hypothesis that would be rejected at the 5% level of significance, while points inside the interval cannot by definition be rejected. Thus, in a test whether the difference between two means is significantly different from 0, if the confidence interval includes 0, then the difference is not significant; but if 0 is not included the difference is significant. Similarly, in testing whether a sample ratio of two rates is significantly different from 1, if a confidence interval includes 1 the difference is not significant; but if 1 is not included the difference is significant.

Another use of confidence intervals is to provide some guidance on whether the sample is too small for reliable conclusions. A very wide interval is generally taken as indicating that the sample is too small. What is too wide is a matter of judgment. The way to narrow an interval is to increase the size of the sample – which of course is not always possible. Here are two examples in which two-sided confidence intervals were required, the first in which sample size could be adjusted, the second in which it could not.

The Commodity Futures Trading Commission, which regulates futures markets, requires them to report on rates of errors in transactions, but permits them to satisfy reporting requirements by using random sampling. The sample must be large enough so that "the estimate can be expected to differ from the actual error percentage by no more than plus or minus [probably two percentage points] when measured at the 95% confidence interval." This specification calls for a two-sided 95% confidence interval of two percentage points above and below the sample estimate. Such an interval consists of all those points that would not be rejected as a null hypothesis at the 2.5% (one-tailed) level of significance given the sample data.

A second example is *Craft v. Vanderbilt*,[4] a case brought in 1994, in which Vanderbilt University was sued for a study on humans conducted almost 50 years earlier. It appears that in 1945, when radioactive tracers were first being used in medical treatment and research, Vanderbilt Medical School conducted an experiment in which some 800 pregnant women visiting a prenatal clinic received a tracer dose of a radioactive isotope of iron. They were told it was a vitamin cocktail. The investigators wanted to study the metabolism of iron in pregnancy and evidently anticipated no harm to the women or their unborn children. Nevertheless, a follow-up study 20 years later revealed that three of the children exposed in utero had died of various childhood cancers when only 0.65 deaths would have been expected based on statewide rates. (A fourth case of liver cancer was disregarded by the investigators because it appeared to be familial.) The relative risk of cancer was thus about 4.6 times greater in the exposed vs. unexposed children. This relative risk was high enough, in the words of the investigators, "to suggest causation." Because there were only three deaths, a 95% confidence interval for the relative risk was from 1.3 to 13.5. Since the lower bound of the interval was greater than 1, the relative risk was statistically significant. But the confidence interval was so wide that a statistician appearing for a defendant could reasonably opine that one could not regard the 4.6 figure as a reliable measure of the association between the trace exposure in utero and childhood cancer. The case was settled before the court passed on the adequacy of the scientific evidence.

Courts sometimes informally recognize that sampling error may vitiate findings based on small samples by applying a "change-one" or even a "change-two" rule: findings are not accepted when the addition or subtraction of a single case or two would substantially affect the percentages shown in the sample. The change-one or change-two rules are ad hoc ways of allowing for sampling error, but the range of results they make are usually much smaller than those indicated by 95% confidence intervals. The cases and the rules are discussed in Chapter 8.

Another reason for computing a confidence interval is that it gives us some insight into the statistical significance of small differences that might be regarded as irrelevant in the law. For example, suppose that a sample study of outcomes of a test given by an employer shows that blacks pass at a rate that is 65% of the white rate. The sample is large enough so that the difference is statistically significant, so we reject the hypothesis that the two rates are equal. But is the sample sufficient to indicate that the black pass rate is less than 80% of the white rate, as the EEOC rule requires for disparate impact? That question would be answered by computing a confidence interval for the ratio of the two rates and seeing whether the upper end of the interval includes 80%. If it does, one would have to conclude that the evi-

---

[4]No. 3-94-0090 (M.D. Tenn. 1994).

dence does not rule out the possibility that the black pass rate is not substantially lower than the white rate by the EEOC's definition. This may not defeat the claim, but could be a factor to consider in deciding whether to require more data.

Confidence intervals can be computed on an exact or approximate basis. Exact computations tend to produce somewhat wider intervals, and so are conservative, but approximate intervals are sufficiently precise, and are commonly used, at least when the sample is not too small. With a sufficiently large sample, the sample mean is approximately normally distributed, so that there is about a 5% probability that it would lie more than 1.96 standard errors from the population mean. Since 95% of the time the sample mean will lie within 1.96 standard errors of the true mean, the sample mean plus or minus 1.96 standard errors creates an interval that with 95% probability will cover the true mean. This is the origin of the common reference to a statistic plus or minus two standard errors. In most cases involving means (as opposed to proportions) the standard error of the statistic is estimated from the sample itself. If samples are small (the cutoff is usually around 30) the t-distribution must be used instead of the normal, which means that a 95% confidence interval will be somewhat wider than plus or minus 1.96 standard errors. When designing a study to meet certain specifications for the confidence interval, as in the Commodity Exchange regulation, the standard error of the statistic may have to be estimated from a separate preliminary study.

A two-standard error approximation cannot be used when a confidence interval is not symmetrical, as is the case for certain binomial distributions, e.g., those with np or nq less than 5. Confidence intervals for rate ratios and odds ratios are also asymmetric. In some of those cases, one proceeds by transforming the data into logs, which makes their distribution more symmetrical, computing a symmetrical confidence interval, and reexpressing the limits in the original form (which usually makes the interval asymmetrical). The log transformation was required to compute the confidence interval for the rate ratio in the Vanderbilt case discussed above.

It is important to remember that a confidence interval does not give a margin for all kinds of possible errors in statistical studies, only sampling error. The U.S. Court of Appeals for the Fifth Circuit got it wrong when in a case involving Bendectin it elevated confidence intervals to a cure-all for confounders in statistical studies. In that case, *Brock v. Merrill Dow Pharmaceuticals*, Inc.,[5] the court had to evaluate epidemiologic studies relating Bendectin taken in pregnancy to birth defects in subsequently born children. The court noted that birth defects might be caused by confounding factors, such as alcohol, nicotine, or viruses, and when epidemiologists compared birth defect rates in women who took Bendectin in pregnancy with those who did not, the results might be skewed by an uneven distribution of such confounding factors between the two groups. So far so good. But then the court jumped to a completely unwarranted conclusion: "Fortunately, we do not have to resolve any of the above questions [as to the existence of confounders], since the studies presented to us incorporate the possibility of these factors by the use of a *confidence*

---

[5] 874 F.2d 307 (5th Cir. 1989).

interval."[6] In this the court was much too sanguine – a confidence interval only measures the range of possible outcomes due to sampling error and offers no other reassurance as to the cause of an observed association in a statistical study. This is obvious from the fact that increasing the size of a sample narrows a confidence interval, but generally does not affect bias due to confounding.

---

[6] *Id.* at 312 (emphasis in original).

Chapter 7
Power

We reject the notion that we are alone in the universe and sweep the skies with powerful radio telescopes, looking for the faintest signal from intelligent life. Since nothing has been heard or received, except the primordial cosmic hiss, is the silence strong evidence that we are indeed alone? The answer, you will say, depends on the probability that if there were intelligent extraterrestrial life our sweeps of the sky would have detected it. This probability is called the power of our procedure. In statistics, a test procedure is said to have high power if it is very likely to detect an effect when one exists; conversely, it has low power if an effect could easily escape detection. Thus, if our test has high power, we are justified in treating the negative outcome as affirmative evidence that there is no extraterrestrial life, but if our test lacks such power, the negative outcome is merely inconclusive.

Our search can have high power against certain possibilities and low power against others. Tests usually have high power against big effects, but lower power against small ones. If we suppose that there is some kind of extraterrestrial life, the power of our test depends on how much and what kind there is. The absence of any signal is probably strong evidence (i.e., the test has high power) against the existence of teeming civilizations like our own on neighboring planets, but only weak evidence (i.e., the test has low power) against the existence of life of any kind in some far-off planetary system. Power also depends on what we proclaim as a positive or negative result. If we insist that the signal be irrefutably sent by intelligent beings, our test will have lower power because we may fail to recognize a communication as such from a civilization very different from our own.

In the search for extraterrestrial life we cannot compute the power of our tests. The matter has too many unknowns. But statisticians routinely do compute the power of their tests in more mundane pursuits. In these more terrestrial contexts power is defined in terms of a null hypothesis, an alternative hypothesis, and so-called Type II error. The null hypothesis, for example, might be an assumption that there is no association between some antecedent factor (e.g., smoking) and an outcome factor (e.g., cancer). If a sufficient association appears in our study data, the null hypothesis is rejected. The alternative hypothesis is that there is a specified association (e.g., smoking increases the risk of cancer by a factor of 10). Type II error is the error of failing to reject the null hypothesis when the alternative is true

M.O. Finkelstein, *Basic Concepts of Probability and Statistics in the Law*, DOI 10.1007/b105519_7, © Springer Science+Business Media, LLC 2009

(e.g., failing to reject the hypothesis of no association between smoking and cancer when in fact smoking does increase the risk of cancer by a factor of 10). The Type II error rate is simply the rate of such errors, or the probability that we would fail to reject the null hypothesis when a specified alternative is true. The complement of the Type II error rate (one minus the Type II error rate) is the probability that we will reject the null hypothesis when the alternative is true; this is the power of the test.

There are no hard and fast rules, but power greater than 80% would generally be characterized as high, while power below 40% would generally be characterized as low. Power of 90% is sometimes regarded as a goal or standard – we would like to be 90% sure of rejecting the null hypothesis when an alternative of interest is true. The theoretical limit for power is of course 100%, a goal that is not realizable in actual tests; at the other extreme, tests with 0% power against certain alternatives do arise in practical settings involving small samples.

The power of a test becomes an issue in a lawsuit when a party seeks to draw inferences from findings of a study that are not statistically significant. The litigation over Bendectin provides an example. Bendectin was a drug that was sold for morning sickness (nausea) in pregnancy. Class actions were brought against Merrill Dow Pharmaceuticals, the manufacturer, on the theory that Bendectin caused birth defects in children exposed in utero. But epidemiological studies of Bendectin and birth defects showed no statistically significantly increased rate of such defects in exposed vs. unexposed children. Nevertheless, plaintiffs sought to have an expert testify, on the basis of non-epidemiological evidence, that Bendectin caused birth defects. Merrill Dow argued that the expert's opinion was scientifically unreliable because epidemiological studies showed no statistically significant elevated risk. Plaintiffs countered that the studies were not evidence of safety because they lacked power – that is, even if Bendectin had in fact increased the risk of certain birth defects, the studies were unlikely to detect it by finding the increase was statistically significant. Thus the power of the epidemiological studies to detect departures from the null hypothesis became an issue in the cases.

To calculate power, conceptually there are three steps. First, one must determine how large a departure from expected values would have to be observed to reject the null hypothesis. Second, one must specify an alternative hypothesis. Finally, one must calculate the probability under the alternative hypothesis of observing such a departure. Let us apply this template to the Bendectin situation.

Researchers studying Bendectin usually proceeded by collecting information from a sample of women who had taken the drug during pregnancy and those who had not. The rates of birth defects in the children of the two groups were then compared. To take one example, the population rate for limb-reduction birth defects is about 1 per 1,000. The null hypothesis being tested is that the rate of such defects is the same for children exposed and unexposed to Bendectin. By the usual standard, we would reject the null hypothesis if the rate of defects among the exposed children in our sample is so much greater than 1 per 1,000 that it is statistically significant at the 5% level. A routine calculation, using the conventional 5% level of statistical significance, gives us the following result: If there were 2,000 exposed children in the study, under the null hypothesis the expected number of limb reductions is 2 and there is less than a 5% chance of seeing 5 or more cases.

The next step is to determine the probability of observing at least 5 limb-reduction cases in a sample of 2,000 exposed children under the alternative hypothesis that there is an increased risk associated with the exposure. What increase shall we assume? One useful candidate is the hypothesis that Bendectin at least doubled the rate of limb-reduction defects, i.e., it became 2 per 1,000 for those exposed. The reason for choosing a doubling is that if the exposure to Bendectin more than doubled the rate of defects in exposed children, then more than half of the exposed children with the defect would not have been defective if there had been no exposure. It follows that it is more likely than not that the limb reduction in a child selected at random from the group of exposed children was caused by the exposure. This logic led the Court of Appeals for the Ninth Circuit to hold, in one of the Bendectin cases,[1] that proof of causation required at least a doubling of the risk. If power is high with regard to this alternative hypothesis, i.e., the null hypothesis is likely to be rejected if Bendectin at least doubled the risk, the failure to reject the null is strong evidence that Bendectin does not have so large an effect, and hence is affirmative evidence against an effect being large enough to be deemed sufficient proof of causation. The fact that power will be lower against smaller effects means that the failure to reject the null may be inconclusive evidence against such effects, but this does not deprive the evidence of its force if smaller effects are deemed insufficient as a matter of law.

If the rate of defects among the exposed children is doubled from 1 to 2 per 1,000, the expected number of defects in the sample would be 4 and the probability of observing 5 or more (at which point we would reject the null hypothesis) is 0.421. We conclude that there is only about a 42% chance that the null hypothesis would be rejected if the alternative we have specified were true. Since the test has rather low power, it is only weak affirmative evidence that the null hypothesis is true, as against that alternative.

Now suppose that the alternative hypothesis is that exposure to Bendectin quadrupled the rate of limb-reduction defects to 4 per 1,000. The expected number of cases in our sample of 2,000 would be 8, and the probability that the number would be at least 5 is about 0.965. Power has increased to 96.5% as against this alternative hypothesis. A nonsignificant finding is thus only weak evidence that the true relative risk is not as high as 2, but very good evidence that the relative risk is not as high as 4. The alternative hypothesis against which power is being measured can thus make a critical difference in the power of the test. For this reason, statisticians frequently speak of a power function, in which power is shown as a curve for different alternative hypotheses. An example of a power curve in studies of the relation between silicone breast implants and connective tissue diseases, a subject of many lawsuits, is shown in Fig. 7.1. The figure shows that the studies, which showed no statistically significant increases in such diseases for women having breast implants,

---

[1] Daubert v. Merrell Dow Pharmaceuticals, Inc., 43 F.3d 1311 (9th Cir. 1995).

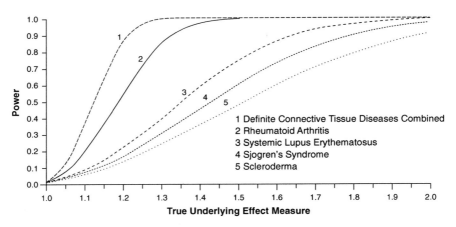

Fig. 7.1 Power vs. true underlying effect measure (relative risk)

had very high power (above 0.9) for relative risks approaching 2 but lower power against smaller relative risks.

In the actual Bendectin cases, the multiple studies of Bendectin, which almost uniformly showed no elevated rate of defects, turned out to have fairly high power against the alternative that Bendectin doubled the rate of defects, and of course even higher power against the alternative that Bendectin quadrupled the rate of defects. The court was thus justified in concluding that the studies were affirmative evidence that the null hypothesis was true. The expert's proposed testimony was treated as conflicting with the epidemiology, and for that reason, among others, it was excluded.

Here is another example. The U.S. Occupational Safety and Health Administration (OSHA) is charged with the task of setting maximum permissible levels of toxic substances in the workplace. Statistical studies are thought to provide important clues to the risks. By regulation, OSHA provides that non-positive epidemiologic studies of potential carcinogens will be considered evidence of safety only if they are "large enough for an increase in cancer incidence of 50% above that in the unexposed controls to have been detected."[2] In other words, evidence of safety requires high power against an alternative hypothesis of a 50% increase. Why measure power against such a large increase? While OSHA admitted that "[a]n excess risk of 50% can hardly be regarded as negligible," it justified the standard by observing that "to require greater sensitivity would place unreasonable demands on the epidemiologic technique." This seems to put the cart before the horse: the excess risk that is important to workers' health should determine which non-positive epidemiologic studies are large enough to be evidence of safety, and not the other way around.

---

[2]OSHA Rules for Identification, Classification, and Regulation of Potential Occupational Carcinogens, 29 C.F.R. § 1990.144(a) (2006).

Another factor that affects the power of a test is the significance level of the test required for rejecting the null hypothesis. The higher the level of significance (i.e., the smaller the P-value) the fewer the rejections of the null hypothesis and the lower the power of the test. For example, if we used a 1% instead of a 5% test, the number of birth defect cases needed to reject the null hypothesis would be increased from 5 to 7 in our example. If Bendecting quadrupled the rate of defects, the probability that the null hypothesis would be rejected is about 69%. Power is substantially reduced from the previous 96% level for the 5% test. This illustrates the tradeoff in test specifications: Requiring a higher level of significance reduces the probability of falsely rejecting the null hypothesis when it is true, but also reduces the probability of correctly rejecting it when it is false. Power and significance are thus inversely related.

The usual way statisticians manage this tradeoff is to select a desired level of significance (commonly 5 or 1%) and then pick the test with the greatest power for that level. However, in some cases power is designated and significance falls out as a consequence. A regulation of the U.S. Environmental Protection Agency (EPA), which was previously discussed in Chapter 5 as an example of the t-test, illustrates the designation of power and the consequences for significance. The regulation specifies that a new fuel or fuel additive must be tested to determine whether it causes vehicles to fail emission standards. The EPA provides that the test specification, applied to a sample of vehicles, must be such that there is a 90% chance of rejecting the new fuel or additive if 25% of the vehicles in the population would fail the test. The Agency thus specifies the alternative hypothesis (25% failure rate) and power (90%), but leaves the number of vehicles to be tested up to the proponent.

In a case involving the new fuel additive called Petrocoal, discussed in Chapter 5, 16 vehicles were tested and 2 failed the test. In order to have 90% power, there must be at least a 90% probability of rejecting the null hypothesis (that Petrocoal does not increase the vehicle failure rate) if 25% of the vehicles in the population would fail the test. For a sample of 16, if the probability of failure for a single vehicle is 0.25, the probability of 2 or more failures is about 93%, and the probability of 3 or more failures is about 80%. Hence, a rejection of Petrocoal must be announced if two or more vehicles fail the test. Under this specification, power is at least 90%, as required. But because the sample is only 16 vehicles, the test specification needed to give this level of power puts the significance level (P-value) of the test far above the conventional 5 or 1% levels. Suppose that the null hypothesis were true and that 5% of vehicles (whether or not driven with Petrocoal) would fail the test. Then the probability of 2 or more failures in 16 vehicles would be about 19%. This is much higher than conventional levels. If the EPA were content with 80% power, our test specification could be raised to 3 or more vehicles and the P-value of the test would become an acceptable 4.3%.

The principal way of increasing power for a test with a given level of significance, or of increasing significance for a test with a given level of power, is to increase sample size. Consider the first Bendectin example in which the sample was 2,000 exposed children, and power was a modest 42% against the alternative hypothesis that Bendectin doubled the rate of birth defects. If the sample is increased to 4,000 exposed children, power rises to about 61%. By doubling the sample size, power has increased about 50%. In the EPA example, the proponents of Petrocoal could have improved the characteristics of the test by increasing the number of vehicles being tested. With 25 vehicles, the test specification that preserves 90% power can be raised to 4 vehicles, while the *P*-value of the test drops from 19 to about 3.4%.

In 1961 in Hot Springs, Arkansas, a black man, William Maxwell, was convicted of raping a white woman and was sentenced to death. In subsequent federal court proceedings attacking the death penalty as discriminatory, a sociologist testified on the basis of a sample study that the probability of a death penalty was far greater if the defendant was black than if he was white.[3] The death-penalty rate was greater for blacks than for whites, but it was argued that factors other than race could have been responsible. The expert argued that such other factors could be ignored because they were not significantly associated with either the sentence or defendant's race.

For example, one argument was that among white defendants there were fewer cases of rape by strangers, which might be viewed by jurors as more heinous. To test this proposition, the sample data included information on authorized and unauthorized entry into the victim's home where the rape occurred. The data showed that the death penalty was meted out to 6 out of 16, or 37.5%, of the defendants whose entry was unauthorized but only to 8 out of 35, or 22.9%, of the defendants whose entry was authorized. However, the difference between the death-penalty rates for the two groups was not statistically significant and on this basis the expert argued that the null hypothesis was true, i.e., there was no association between this factor and the sentence.

What the expert evidently omitted from his analysis was the low power of his test of significance. A natural alternative hypothesis is that the difference in the two rates is that observed in the data, or 14.6 percentage points. If this alternative were true, the probability of rejecting the null hypothesis with a 5% one-tailed test is only about 20%. Power is thus very low, meaning in this context that the absence of significance is not strong evidence that the factor of unauthorized entry is unassociated with the death penalty. Subsequent studies of the death penalty have measured the racial effect after taking such factors into account. See p. 159.

When a test generates results that are in the suspected direction, but are not statistically significant, and power is low, the evidence is inconclusive. In some cases, however, the data should be viewed as suggestive. The EEOC four-fifths rule is a case in point. The rule sensibly suggests that if any race, sex, or ethnic group passes an employment test at less than 80% of the rate for the group with the highest pass rate, but the difference is not statistically significant, one should collect data on the impact of the procedure over a longer period or with other employers. In effect, the EEOC recognizes that the power of the test in such cases may be low.

---

[3]Maxwell v. Bishop, 398 F.2d 138 (8th Cir. 1968), *vacated and remanded*, 398 U.S. 262 (1970).

Chapter 8
Sampling

What is Sampling?

We all know what sampling is: In very general terms it has been defined as "methods for selecting and observing a part (sample) of the population in order to make inferences about the whole."[1] That seems noncontroversial enough, but what is encompassed by the term "sampling" in a statute has been the subject of litigation in the U.S. Supreme Court.

The case arose when Utah sued the U.S. Department of Commerce to stop the Census Bureau from "imputing" demographic information to households on the master address list that did not return the census questionnaire and for which good information could not subsequently be obtained by follow-up procedures.[2] For each unsuccessfully enumerated address, the Bureau imputed population data by copying corresponding data from a "donor" address. The donor address was the "geographically closest neighbor of the same type (i.e., apartment or single-family dwelling) that did not return a questionnaire by mail." In the 2000 census imputation added about 1.2 million people, representing about 0.4% of the entire population.

The Census Act prohibits "sampling" for determining state populations for reapportionment purposes. Utah, which lost a representative because of imputation, objected that imputation was a form of sampling. By the broad definition, Utah had a point: The neighbor whose values are imputed to the missing household is part of the population who is selected and observed in order to make inferences about the whole. But Justice Breyer, writing for the majority, held otherwise.

Breyer's opinion argues that the *nature, methodology*, and *immediate objective* of imputation distinguished it from sampling. The nature of the Bureau's enterprise was not the extrapolation of features of a large population from a sample, but the filling in of missing data as part of an effort to count individuals one by one. Nor did the methodology involve a random selection of donor units from which

---

[1] Kish, Survey Sampling 18 (1965).

[2] Utah v. Evans, 536 U.S. 452 (2002).

M.O. Finkelstein, *Basic Concepts of Probability and Statistics in the Law*, DOI 10.1007/b105519_8, © Springer Science+Business Media, LLC 2009

inferences were to be made about unenumerated addresses. Finally, the Bureau's immediate objective was to fill in missing data, not extrapolate the characteristics of the donor units to an entire population. These differences, plus legislative history (Congress evidently did not have imputation in mind when it prohibited sampling for reapportionment purposes) and testimony of Bureau statisticians that Bureau imputation was not "sampling" as that term was used in the field of statistics, convinced him to hold that imputation was not prohibited.

Breyer's line-drawing was made more difficult because the House of Representatives had previously sued to stop the Bureau's plan to adjust the census for the undercount by post-census sampling and had won.[3] The Bureau's program included a plan to adjust for unreturned questionnaires in a census tract by taking a random sample of the addresses, intensively investigating those, and extrapolating the results to the total nonresponse in the tract. The Court held that this would be prohibited sampling. Referring to the House of Representatives case, Breyer conceded that the prohibited program also resembled an effort to fill in missing data because the sample of nonrespondents would be used by the Bureau to extrapolate values for only about 10% of the tract. Nevertheless Breyer found key differences in the deliberate decision taken in advance to find an appropriate sample, use of random sampling to do so, the immediate objective of determining through extrapolation the characteristics of the entire nonresponding population, and the quantitative figures at issue (10% there; 0.4% here). These differences, taken together, distinguished sampling in that case – "in degree if not in kind" he admitted – from the imputation at issue.

Justice O'Connor disagreed. Among other things, she pointed out, correctly, that sampling was not confined to estimates of large populations from small samples or to random samples. And focusing on the Bureau's "immediate objective" of filling in missing data overlooked the fact that the Bureau estimated nonresponse population characteristics using a selected subset of the population and imputation was simply a means to that end. It is hard to deny these points. Perhaps the best argument for Breyer's result is the one based on legislative history: The prohibition was intended to be confined to sampling in the more traditional sense. Thus, depending on the context, sampling in law may have a narrower definition than the broadest definitions used by statisticians.

Simple Random Sampling

There have been many legal proceedings in which the result turned on information obtained from samples and extrapolated to the populations from which the samples were selected. From the statistician's point of view, the validity of such extrapolations does not turn – as is commonly supposed – on the size of the sample relative to the size of the population, but primarily on the way in which the sample was selected

---

[3] Department of Commerce v. United States House of Representatives, 525 U.S. 316 (1999).

and analyzed. Briefly describing such ways and their limitations is the purpose of this chapter.

Sampling plans come in many varieties and degrees of complexity, but the easiest to understand is the *simple random sample*, in which, as each member of the sample is selected, each member of the population has the same chance of being included in the sample and the selection is made by some objective mechanism that ensures that result. The gold standard is a case in which it is possible to number the population (the set of these numbers being called the *sampling frame*) and then to select a sample using a pseudo-random number generator included in most statistical computer programs. In such samples, the difference between the sample estimate and the population parameter being estimated is due to two factors: (1) bias, or systematic error, and (2) random error, or chance. Thus *sample estimate = population parameter + bias + random error*. Bias can in theory be eliminated, although this is very hard to do in all but the simplest cases. On the other hand, random error (also called *sampling error*) cannot be eliminated, but at least can be estimated.

Physical randomization, such as drawing labeled balls from a well-mixed urn, can also be used, but is less reliable. More than a casual effort has to be made to secure randomness. An example is the 1970 draft lottery for the Vietnam War. In the lottery, men were called up by order of birth date. The order was determined by putting slips for the days of the year into a bowl and selecting slips at random until all 366 slips were picked. Order mattered because Selective Service officials announced that those in the top third would be called, those in the middle third would probably be called, and those in the bottom third were unlikely to be called. The slips were put into the bowl by date order, beginning with January and ending with December. Despite the fact that the bowl was stirred, statistical analysis showed that later dates (which were put in last) tended to be selected early and earlier dates were selected later. Evidently the stirring was not good enough to create a truly random ordering. To correct for this bias, the 1971 draft used two bowls: in one were slips for the birth dates and in the other were slips with the selection numbers. A date drawn from one bowl was given a selection number by drawing from the other bowl. Tests showed that this eliminated the bias.

A third method that is commonly used when the population can be put on a list is to pick every *n*th member, with a random starting point selected by picking a random number from the first *n* members on the list. Because the starting point is random, each member of the population has the same probability of being included in the sample. If there is periodicity in the list the sample will not be biased (because the random starting point gives each member the same chance of being included) but the sampling error will be larger than is shown by the usual computations.

Such methods are called *probability sampling* and the samples selected in one of these ways are called *random samples* or *probability samples*. Such random samples have to be distinguished from *haphazard* or *convenience samples*, in which the choice of the sample is to some degree left to the discretion of the collectors. This is true in *quota sampling*, in which the collectors have to fill quotas of interviews within specified groups, e.g., white males 19–25, but are otherwise free to pick the respondents. Quota sampling is not random sampling and may introduce bias.

The purpose of random sampling by an objective method is to (1) avoid bias in the selection and (2) allow statisticians to estimate random error by computations of confidence intervals for sample estimates. It is sometimes also said that the purpose is to secure *representative* samples. The term has no objective or technical meaning, so whether a sample is representative is a matter of judgment (if it can be known at all), not mathematics. It is true that in general, with sufficiently large samples, the sample estimates will tend to be close to the population values, but it has to be remembered that the term is vague and there can be bad breaks in random sampling.

The Problem of Bias

In simple random sampling chance error is well understood mathematically and can be estimated. The principal determinants of sampling error have already been discussed. See Chapter 3. Bias is much harder to deal with and frequently cannot be measured; as a result it has received less attention. Bias comes in three principal varieties: (i) selection bias; (ii) nonresponse bias; and (iii) response bias.

To illustrate *selection bias*, the favorite example of statisticians is the 1936 *Literary Digest* poll, which predicted that Landon would beat Roosevelt in the presidential election of that year. The prediction was based on a poll in which questionnaires were sent to 10 million people, of whom 2.4 million replied, the largest ever taken. But it was wildly wrong because the names and addresses of those people came from sources like telephone books and club membership lists that were biased against poorer people (in those days only one household in four had a telephone) and those people voted disproportionately for Roosevelt.

In a case involving household interviews, a sugar company, Amstar Corp., sued Domino's Pizza, Inc., claiming that the pizza company's use of the name "Domino" on its pizza infringed the Domino Sugar trademark, which was owned by Amstar. The legal question was whether buyers of pizza thought that the pizza company also made sugar. Amstar surveyed 525 persons in 10 cities (only two of which had Domino's Pizza outlets). The persons interviewed were women reached at home during the day who identified themselves as the household member responsible for grocery buying. Shown a Domino's Pizza box, 44.2% of those interviewed thought that the company that made the pizza also made other products; 72% of that group (31.6% of all respondents) believed that the pizza company also made sugar. The court of appeals rejected the study as proving consumer confusion, in part because the women who were interviewed were not the pizza buyers in the household and in all but two of the cities no one in the household could have bought a Domino's pizza.[4]

Selection bias can be an issue in statistical studies of the legal system that have become increasingly prominent in legal academia. For example, a major study of

---

[4] Amstar Corp. v. Domino's Pizza, Inc., 615 F.2d 252 (5th Cir. 1980).

death-penalty cases, by Professor James Liebman of Columbia Law School and his colleagues, reviewed over 5,800 such sentences imposed between 1973 and 1995. The study made national headlines when it reported that the reversal rate in such cases was a shocking 68%. Most reversals were caused by egregiously bad defense lawyering or prosecutorial misconduct. There was, however, a problem that put a question mark over that finding: a significant number of cases were still pending when the study window closed in 1995. In statistical parlance such cases are known as *censored* because an intervening event (here, the end of the study) keeps their outcome from being known. The investigators considered what to do and decided to exclude censored cases. This had the merit of simplicity and transparency but created a risk of bias: the principal concern was that if cases with longer durations or more recent sentences had lower reversal rates, the 68% figure could be biased upward by the exclusion of censored cases.

Statisticians usually try to correct for such biases by either (i) using the observed cases to estimate the outcomes for the censored cases or (ii) weighting the outcomes of the observed cases to eliminate bias in the statistics based on them. Here's how method (i) works as applied to the Liebman study. We begin by arraying all the censored cases by starting year. Suppose the first group of such cases had 1980 as their starting year. By 1995 when the study closed they were still pending after 15 years. To estimate the probability of reversal after 15 years for cases in this group, we look at the reversal rate for a class of "reference" cases: these are cases with start dates prior to 1980 that lasted more than 15 years. All these reference cases are observed because 1980 is the earliest start date for a censored case. We then apply that reversal rate to the 1980 censored cases to impute the number of reversals for that group. In effect, this approach assumes that, with respect to outcome, the censored cases were selected at random from all cases with durations of more than 15 years. We then proceed to the next oldest group of censored cases. Suppose they had a start date in 1985 and thus were still pending 10 years later in 1995 when the study closed. To estimate the probability of reversal for this group, we look at reference cases with start dates prior to 1985 that lasted more than 10 years. Some of these cases will be observed and some will be censored (the 1980 group). The reversal rate for the 1985 group is the number of observed reversals plus the number imputed at the first step for the 1980 group, divided by the total number of reference cases. This rate is then applied to the 1985 group to estimate the number of its reversals. The estimate for the 1985 group thus builds on the estimate for the prior group. We proceed in this fashion for each successive group until estimates have been made for all groups of censored cases. We can then estimate the probability of reversal for all cases as the number of reversals for the observed cases plus the number of imputed reversals for the censored cases, divided by the total number of cases, observed and censored.

The other method for dealing with censored data is, as we have said, to weight the outcomes of the observed cases in such a way as to reduce or eliminate the bias in estimates based on them. In general, when some observations in a data set are missing, the weight statisticians would assign to an observed case is the reciprocal of the probability of its being observed, that is, the probability that it would not be missing. The logic is this: Suppose a given case outcome has a 20% probability of being observed. Thus for this case, and every other one like it that is observed, there will be four others on average that are missing. Therefore, the observed case should "stand in" for itself and these four others, so it should be weighted by a factor of $1/0.2 = 5$. The probability of being observed for each censoring time is estimated from the observed cases. In the death-penalty reversal rate study discussed above the two methods lead to the same result.[5]

Methods such as these may be challenged because they usually involve assumptions about the missing data that cannot be verified. In our example, the key assumption was time homogeneity, that is, the duration of the case and the probability of reversal were statistically independent of start date. When assumptions like this are questionable, there are modifications that statisticians may use to address the problem, but in any event taking account of censored cases, as we have described, even if the assumptions of the method are not fully satisfied, should generally lead to a less biased description than simply ignoring them. In the death-penalty study, when we adjusted the data to account for the censored cases, instead of just excluding them, the reversal rate dropped to 62%, indicating that there was some bias but not enough to affect the basic thrust of the study.[6]

Missing information and nonresponse are endemic problems in samples and surveys. If the nonrespondents are different from those responding in ways that matter, the sample results will be biased. In the *Literary Digest* poll described above, the *Digest* made a special survey in which questionnaires were mailed to every third registered voter in Chicago. About 20% responded and of those responding over half favored Landon. But in the election, Chicago went for Roosevelt by a two-to-one majority. Evidently, nonresponse was a major problem in the special survey.

One way to test whether nonresponse is a problem is to compare responders and nonresponders with respect to some quantity (a "covariate") that is known for both, or to compare the responders with some covariate known for the population. If the responders' covariate values are close to those for nonresponders or to the population, that is some evidence that nonresponse or missing information has not skewed the sample. Here are two examples.

A mail-order house in Colorado sold stationary and novelty items in California. California claimed a tax on such sales because the company had "a substantial presence" in the state. The company responded that part of its sales were at wholesale for which no tax was due. To substantiate its claim, the company made a sample survey of its mail orders by sending a questionnaire to every 35th California order. The response rate was about 60%. The company treated the sample response as representative of all its orders. This could be tested because the distribution of orders by the covariates of dollar size and city was known for both the responders and the

---

[5]For a more detailed description of these methods and their application to the Liebman study, see M.O. Finkelstein, B. Levin, I.W. McKeague & W-Y. Tsai, "A Note on the Censoring Problem in Empirical Case-Outcome Studies," 3 Journal of Empirical Legal Studies 375 (2006).
[6]*Id.* at 382.

nonresponders. The evidence of these covariates was mixed: There were proportionally fewer responders than nonresponders for low-priced orders and the difference was significantly different; but the proportions of orders from San Francisco, Los Angeles, and "Other" cities were virtually identical for both responders and nonresponders. The court accepted the sample as "representative."

In *Rosado v. Wyman*,[7] due to the passage of time, only 62.6% of the welfare records in a random sample could be found; the court nevertheless accepted the sample after noting that the average payment and family size approximated those known characteristics of the whole population. But in *E.E.O.C. v. Eagle Iron Works*,[8] a discrimination case, data for 60% of current and former employees were rejected as unrepresentative where all the missing racial data were from former employees.

Many studies have preplanned sample sizes and if a nonrespondent is encountered a respondent is substituted. This may reduce sampling error but does not reduce bias due to nonresponse. To address that, the number of nonresponders must be known.

Response bias arises whenever the information collected from respondents is sufficiently inaccurate to skew the results. This may occur because of faulty recollection or the respondent's bias or because the subject matter is sensitive. For example, women who had received silicone breast implants sued the manufacturers claiming that silicone leaking from the implants had caused rheumatoid arthritis, lupus, and other auto-immune connective tissue diseases (CTD). A federal district judge appointed an expert panel to appraise the scientific issues, including epidemiologic studies of the connection. The panel examined 16 such studies, of which 15 showed no heightened incidence of CTD for women with implants. One study (the largest) did show an increase, but this was discounted by the authors of the study because it relied on self-reporting by the women as opposed to hospital records, and at the time the women reported there was much publicity surrounding claims that the implants caused CTD. On the basis of the epidemiologic studies and other evidence, the panel reported to the judge that no connection was shown, and this led to the drying up of the flow of new cases and the settlement of most pending cases.[9]

More Complex Designs

In sampling from a large and diverse population it is almost always necessary to use more complex designs than simple random sampling from the entire population. The principal additional techniques are *stratified random sampling* and *cluster sampling*. In stratified random sampling, the population is divided into strata, separate random samples are taken from each stratum, and the results are combined.

---

[7] 322 F. Supp. 1173 (E.D.N.Y. 1970).

[8] 424 F. Supp. 240 (S.D. Ia. 1976).

[9] B.A. Diamond et al., "Silicone Breast Implants in Relation to Connective Tissue Disease..." avail. at www.fjc.gov/breimlit/md1926.htm (1998).

Ideally, the strata are more homogeneous than the population as a whole, so each random sample has a smaller sampling error than would be the case for a random sample from the whole population. Stratified sampling also ensures that each stratum is properly represented in the sample. For example, each month the Census Bureau makes national estimates of employment and unemployment by means of a sample survey called the Current Population Survey. The Bureau divides the country into strata (almost 800 of them) consisting of cities or counties that resemble each other on certain demographic characteristics and so are presumably more uniform than the country as a whole; it then samples within those strata. The results are then combined to give national and regional unemployment figures.

It is frequently the case in strata sampling that different strata will be sampled at different rates to reduce sampling error. Accountants sampling invoices in an audit will sample larger invoices at higher rates (sometimes 100%) than smaller ones. In an experiment to determine the effect of small classes on student achievement, inner city schools were over-represented in the sample of schools that were given such classes. When this is done, analytic procedures must take the different sampling rates into account in estimating an overall figure. One method for doing so is to weight each unit by the inverse of the probability that it would get into the sample. This estimator is unbiased, but its variance may be high.

In *cluster sampling* a random sample is taken of special units. The special units may be quite large, such as cities or as small as city blocks. Subsamples are then taken from the special units. Cluster sampling was used in a landmark case in which Zippo Manufacturing Company sued Rogers Imports, Inc., claiming that Rogers lighters were confused with Zippo lighters. The universe was adult smokers in the continental United States (then 115 million people). Three separate surveys of approximately 500 persons each were selected using a cluster sample design. The court described the procedure in this way: "The samples were chosen on the basis of data obtained from the Bureau of Census by a procedure which started with the selection of fifty-three localities (metropolitan areas and non-metropolitan counties), and proceeded to a selection of 100 clusters within each of these localities – each cluster consisting of about 150–250 dwelling units – and then to approximately 500 respondents within the clusters."[10]

Because members of clusters tend to be similar, cluster sampling generally has a larger sampling error than an independent sample of the same size. Cluster sampling is therefore not a way of reducing sampling error as stratified sampling is, but of reducing cost: Contacting possibly far-flung members of a random sample from a large population is much more expensive than contacting nearby members of a cluster.

Another technique for reducing sampling error is called *ratio sampling*. To illustrate the method, the Bureau of the Census uses ratio estimation in the Current Population Survey mentioned above. Suppose in one month the Bureau took a sample of 100,000 people of whom 4,300 were unemployed. Since the sample was about 1

---

[10]Zippo Manufacturing Co. v. Rogers Imports, Inc., 216 F. Supp. 670, 681 (S.D.N.Y. 1963).

in 2,000 people in the total adult, noninstitutional population, the Bureau could estimate a national unemployment figure as 4,300 × 2,000 = 6,800,000. However, the Bureau breaks down the sample into segments (e.g., white men aged 16–19) and, in effect, computes an unemployment rate for each segment from the sample, and then weights that rate by the proportion of the segment in the total population. The population proportion for each segment is taken, with adjustments, from the census, not from the sample. The weighted sum of unemployment rates for all groups is the national unemployment rate. The advantage of the ratio procedure lies in the fact that because the number of unemployed in each segment and the size of the segment are likely to be correlated, their ratio (the unemployment rate for the segment) is likely to vary less than each figure taken separately; use of the ratio will thus reduce the sampling error of the estimate.

Complex, multistage sampling plans often make it impossible to provide formulas for sampling variability. One technique to measure such variability involves splitting the sample in various ways and computing the sample results for each of the subsamples. The variation in results provides a direct assessment of variability, which is usually expressed as the root mean squared difference between each subsample and the average of the subsamples. These techniques have the advantage of measuring all variability, both sampling and nonsampling.

Sample surveys appear most frequently in trademark cases and there is much law on the question whether defects in the surveys should make them inadmissible or merely go to their weight. A much-followed statement of the requirements for surveys was given by Judge Glasser in *Toys "R" Us, Inc. v. Canarsie Kiddie Shop, Inc.*[11]:

> The trustworthiness of surveys depends upon foundation evidence that (1) the "universe" was properly defined, (2) a representative sample of that universe was selected, (3) the questions to be asked of interviewees were framed in a clear, precise and non-leading manner, (4) sound interview procedures were followed by competent interviewers who had no knowledge of the litigation or the purpose for which the survey was conducted, (5) the data gathered was accurately reported, (6) the data was analyzed in accordance with accepted statistical principles and (7) objectivity of the entire process was assured.

Not many surveys made for litigation pass those exacting standards.

Small Samples

How big should a sample be? This is a commonly asked question for those designing studies. Lawyers don't usually get involved at the design stage, but after the fact. For them the question is, When is a sample too small? There is no simple answer to this question. In civil rights cases, the courts have rejected small samples, even when they show statistically significant differences, and when they do they quote from the Supreme Court's comment that "[c]onsiderations such as small

[11] 559 F. Supp. 1189, 1205 (E.D.N.Y. 1983).

sample size may of course detract from the value of such [statistical] evidence."[12] True enough, but when a sample crosses the line and becomes too small remains a mystery. Sometimes the courts apply a rubric that if changing the result in one or two cases significantly changes the percentage comparisons the sample is not sufficient (the "change-one" or "change-two" rule). This just moves the subjectivity back a notch (what is a significant change?). If the change-one rule is intended to reflect the fact that in another sample of data the results could by chance be quite different, then calculation of a confidence interval is the accepted way to express that uncertainty and test for statistical significance. In most cases a 95% confidence interval will be much wider than the variation associated with a change-one or even a change-two rule.

For example, in one case a black employee was discharged for illegal acts outside of employment. He sued, claiming that the company's policy in that regard had a disparate impact on blacks. An expert examined personnel data for 100 employees who had been discharged for dishonest acts; of these there were 18, of whom 6 were black, who had committed dishonest acts outside of employment. Blacks thus constituted 33% of those so discharged; by contrast the workforce was 4.6% black. The appellate court found the sample too small to support a finding of disparate impact because the subtraction of even one or two blacks would change the percentages "significantly."[13] By contrast, a 95% confidence interval for the percentage black among those discharged for illegal acts outside of employment is from about 12 to 55%, which is about equivalent to changing four blacks to whites. Since even the lower bound of the confidence interval is well above the 4.6% blacks in the workforce, the difference is statistically significant and it is hard to see why the court found the effect of changing one or two so important.

Recognizing that a change-one rule is a crude way of allowing for chance variation makes it clear that it is overcorrecting, as one court did, to imagine one or two cases changed in result (to allow for chance variation) and then with the new percentages to test for significance (to allow for chance again).[14]

This is not to say that a small sample that is statistically significant is always sufficient; there are contexts in which small samples are statistically significant but are insufficient because of their small size. For example, in one case, orthodox Jewish employees, who were terminated, sued their employer, a Jewish organization, claiming discrimination against the orthodox. There were 47 employees, of whom 5 were orthodox Jews. Three orthodox Jews and four other employees were terminated. The difference in termination rates for orthodox (60%) and non-orthodox employees (9.5%) was highly statistically significant ($P = 0.012$). It was thus unlikely that so many orthodox would have been picked by chance, assuming that they were no

---

[12] Int'l Brotherhood of Teamsters v. United States, 431 U.S. 324, 340, n.20 (1977).

[13] See, e.g., Oliver v. Pacific Bell Northwest Tel. Co., 106 Wash. 2d 675 (1986).

[14] Waisome v. Port Authority of New York and New Jessey 948 F.2d 1370 (2d Cir. 1991) (a written examination did not have a disparate impact on blacks because the disparity in pass rates would lose statistical significance if two additional blacks had passed the test).

more likely to be terminated than non-orthodox employees. But since only three orthodox were involved, it could not convincingly be said, on the basis of the statistics alone, that their terminations were based on some religious factors common to the orthodox as opposed to valid individual factors. Hence, the lopsided distribution of terminations, although statistically significant, was not necessarily legally significant. In refusing to dismiss the case, the court did not rely on statistics, but on statements by company officers indicating an animus against the orthodox.

Sampling has been generally accepted by the courts, but not when it conflicts with strong traditions reflected in constitutional provisions or statutes. We have already seen that sampling is prohibited by statute for the census. Mass torts are another example. In some mass tort cases, judges have attempted to handle the tidal wave of litigation by trying random samples of cases and applying the results to the entire class. District Judge Robert Parker sensibly did this for the damage phase of several thousand cases in which plaintiffs sued a manufacturer and supplier for personal injury and wrongful death caused by exposure to asbestos. Separate random samples of between 15 and 50 cases were selected for each of 5 disease categories. In each category, the sampled cases were tried as to damages before juries, the damage awards were averaged, and then applied to the remaining cases in the category. After the trials, Judge Parker held a special hearing on whether the sampled cases were representative of the cases in their categories. A statistician testified that covariates of the samples and the populations were within confidence intervals of each other. The defendants had data on the populations and the samples but presented no conflicting analysis. The court concluded that the samples were indeed representative and entered judgment on the verdicts. On appeal, however, the court of appeals held that the extrapolation denied the defendants their constitutional and state law rights to an individualized jury determination of damages.[15] Despite the sensibleness of the approach, no appellate court has yet approved sampling as a way of handling the extraordinary problems of mass tort litigation.

---

[15]Cimino v. Raymark Industries, Inc., 151 F.3d 297 (5th Cir. 1998); for a discussion of the district court's approach see M.J. Saks & P.D. Blanck, "Justice Improved: The Unrecognized Benefits of Aggregation and Sampling in the Trial of Mass Torts," 44 Stanford L. Rev. 815 (1992).

Chapter 9
Epidemiology

Epidemiology is the study of the distribution and causes of disease in mankind. It makes use of what are called observational studies, so named because they do not involve creating experimental data, but rather assembling and examining what data exist. This is usually done by comparing rates of disease in exposed and unexposed populations. Observational studies thus take advantage of natural variation in exposures to tease out or test causal relations. Below I describe in broad outline the methods used in two important kinds of epidemiological studies that have appeared in law cases and some of the key issues that arise in making causal inferences from such studies.

Cohort Studies

The most reliable epidemiological investigations are *cohort studies*, in which "exposed" and "unexposed" groups are defined and then followed and compared with regard to the incidence of disease. The comparison is usually made by computing a relative risk, viz., the rate of disease in the exposed group divided by the rate in the unexposed group. Alternatively, the rate of disease in the exposed group may be compared with expected rate in that group based on disease rates drawn from the general population. A relative risk that is statistically significantly greater than one indicates that there is an association between the exposure and the disease. This association – if not the product of bias, confounding, or chance – is evidence that the exposure causes the disease, or more specifically, that the excess number of cases in the exposed group was caused by the exposure. Such an inference, referred to as *general causation*, is more likely to be accepted when the study satisfies certain requirements, known as the Bradford-Hill criteria. These are discussed below. In a lawsuit when causation is an issue, there is the further question whether the plaintiff is one of those in the "base" subgroup, who would have gotten the disease in any event, or is in the "excess" subgroup, whose disease was caused by the exposure. In law, this is called *specific causation*. In epidemiology, it is called the *attributable risk* in the exposed population. The attributable risk is the proportion that the excess subgroup bears to the whole exposed and diseased group. It is thus the proportion

M.O. Finkelstein, *Basic Concepts of Probability and Statistics in the Law*,
DOI 10.1007/b105519_9, © Springer Science+Business Media, LLC 2009

of those exposed people with the disease who would have avoided it if the exposure had been eliminated. It is also the probability that a randomly selected exposed and diseased person would be in the excess group.

Simple algebra shows us that the attributable risk is equal to $1 - 1/$(relative risk). As the formula shows, if the relative risk is greater than two, then for a randomly selected plaintiff it is more likely than not that the disease was caused by the exposure. Whether this should be sufficient for a prima facie case of specific causation has been debated by legal scholars. When the relative risk is between one and two, specific evidence obviously is required for a finding that a plaintiff is in the excess subgroup. If the relative risk is not significantly different from one, which indicates no association, does the negative epidemiology "trump" other evidence? As we noted earlier, if epidemiology has high power and is negative, it has been held to preclude specific expert testimony that is offered to prove causation.[1]

Here are two examples of cohort studies. In the 1950s the federal government conducted above ground tests of atomic weapons in the Nevada desert. Radioactive fallout drifted east into Utah, with the result that 17 counties in southern and eastern Utah were designated as "high-fallout" counties on government maps. Suits were brought against the government for cancer deaths allegedly caused by the fallout. Epidemiologic studies of leukemia deaths in children were conducted using high- and low-exposure cohorts. The high-exposure cohort consisted of children living in high-fallout counties who were aged 14 or under during the period of the tests. The low-exposure cohort consisted of children living in those counties who were born either sufficiently before or after the tests so they were not aged 14 or under during the period of the tests. Deaths of children at age 14 or under, from leukemia, were then counted for the two cohorts. For each cohort, the death rate was computed by dividing the number of deaths by the person-years in the cohort. (For each year, person-years in each cohort was the number of persons who, if they had died in that year, would have had their deaths assigned to that cohort for that year.) The leukemia death rate for the high-exposure cohort was 4.42 per person-year and for the low-exposure cohort was 2.18. The relative risk was $4.42/2.18 = 2.03$. Thus, if one credits the study, slightly more than 50% of childhood deaths from leukemia in the high-exposure cohort were caused by the government's tests. The district court held the government liable for all leukemia deaths on the ground that the government created "a material increased risk." On appeal, however, the judgment was reversed on the ground that the testing program was a discretionary governmental function for which suit was not permitted under the Federal Tort Claims Act.[2] Congress subsequently voted compensation.

In a small community near Los Angeles, parents of children who had died of neuroblastoma, an extremely rare form of embryonic cancer, sued a rocket fuel

---

[1] See p. 16.

[2] Allen v. United States, 588 F. Supp. 247 (D. Utah 1984), *rev'd*, 816 F. 2d 1417 (10th Cir. 1987), *cert. denied*, 108 S. Ct. 694 (1988).

manufacturing company, Aerojet-General Corp., contending that their children's deaths were caused by carcinogenic wastes released by the company into the environment.[3] Plaintiffs' experts made an epidemiologic study comparing the cluster of the four neuroblastoma cases found in Caucasian children in the community with controls consisting of rates of such cases in groups of unexposed counties. The experts searched for control counties by taking the 421 U.S. counties for which cancer data were available from the federal cancer registry (so called SEER counties) and picking counties that matched ($\pm$ 3%) the exposed county on the demographic factors of gender, age, race, educational attainment, income above the poverty level, and median household income. But there were no such counties, so the investigators had to relax the matching criteria of ethnicity and median household income to find controls. Which controls were used made some difference: The relative risks using different groups of control counties ranged from 4.47 to 12.52, and in every instance they were statistically significant.[4] The case settled for $25 million.

Case–Control Studies

Another design is the *case–control* study. In such a study, the cases are a sample of persons with the disease and the controls are another sample of persons who are like the cases, except that they do not have the disease. The exposure history of each group is then compared. The reason for using a case–control study is that, if the disease is rare, very large samples of exposed and unexposed persons would have to be followed as cohorts to get enough cases to make reliable estimates of rate ratios. By beginning with persons who have the disease, a case–control study is more efficient in assembling data. But there are two problems with case–control studies: interpreting the results and finding good control groups.

From case–control samples one can compute the retrospective relative risk of exposure, i.e., the rate of exposure for those who have the disease divided by the rate of exposure for those who do not. This relative risk of exposure is retrospective because we look backward in the causal sequence from disease to exposure. In the usual case, however, our interest is in the relative risk of disease given exposure, which is in the prospective sense, looking forward from cause to effect. Because of its design, which begins with the selection of disease cases and undiseased controls, the case–control study does not permit us to compute prospective statistics involving the probability of disease. The numbers of diseased and undiseased persons are determined by the design of the study and do not fall out from the exposure. However, we can compute a retrospective odds ratio, e.g., the odds on exposure given

---

[3] Pennington vs. Aerojet-General Corp., Case No. 00SS02622, Cal. Superior Ct., Sacramento County.

[4] A description of the epidemiologic studies in the case and the response of defendants' experts may be found in F. Dominici, S. Kramer & A. Zambelli-Weiner, "The Role of Epidemiology in the Law: A Toxic Tort Litigation Case," 7 Law, Probability and Risk 15 (2008).

the disease divided by the odds on exposure given no disease. This retrospective statistic would also be of little interest, except for the convenient mathematical fact that the retrospective odds ratio is equal to the prospective odds ratio: The odds ratio for exposure given disease and no disease is equal to the odds ratio for disease given exposure and no exposure. This is nearly what we want. The final step is this: as noted on pp. 39–40, when the disease is rare in both the exposed and unexposed populations (and in most case–control studies it is), the prospective odds ratio is approximately equal to the prospective relative risk, and it is this statistic that, as we have seen, is the basis for inferences of causation.

The basic problem in a case–control study is to select controls that match the cases in everything relevant to the probability of disease, except that the selection must be made without regard to exposure history. This problem raises issues of bias and confounding that I will address. Here are two examples of case–control studies in which such issues could be raised.

When it was believed that an intrauterine contraceptive device called the Dalkon Shield might cause pelvic inflammatory disease (PID), many lawsuits were brought and case–control studies made of the connection. In one such study, the cases and controls were samples of women who were discharged from 16 hospitals in nine U.S. cities from October 1976 through August 1978. The cases were women who had a discharge diagnosis of PID; the controls were women who had a discharge diagnosis of a non-gynecological condition. Excluded were cases and controls who had a history of PID or conditions that might lower the risk of pregnancy. The studies strongly supported an association. The odds on wearing a Shield vs. another IUD for women who had PID were 0.23, but for women who did not have PID were 0.045. The odds ratio was 0.23/0.045, or about 5. Thus the odds of PID for a woman wearing the Shield were about five times greater than for a woman wearing another IUD. This is a strong result, suggesting causation. However, there is a question (discussed in the next section) whether the cases and controls were properly matched.

In 1983 a British television program reported a cluster of five leukemia deaths among young people in Seascale, a tiny village about 3 km south of the Sellafield nuclear processing plant in West Cumbria in northwest England. Based on national rates, only 0.5 such deaths would have been expected. Many Seascale men worked at Sellafield. Professor Martin Gardner proposed an explanation: Preconception paternal irradiation (ppi) of male workers at the Sellafield plant caused mutations in their spermatagonia which, in turn, caused a predisposition to leukemia and/or non-Hodgkins lymphoma in their offspring. Lawsuits were brought against the nuclear facility on this theory.

In case–control studies to test the Gardner hypothesis as it applied to leukemia, the subjects were 46 leukemia deaths occurring in young people born and diagnosed in the West Cumbria health district in the period 1950–1985. After data collection began, an exception was made and a young man was included as a case who was born in Seascale, but who was diagnosed with leukemia in Bristol, where he was attending the university. There were two groups of controls matched to each case: area and local. For the area controls, searches were made backward and forward in time from the case's entry in the West Cumbria birth register until the nearest four appropriate controls of the same sex in each direction were found. For the local controls, the residence of their mothers was matched for residence (civil parish) of the case; otherwise the procedure was the same as for the area controls. As an example of exclusions: eight potential controls for the young man diagnosed in Bristol were excluded, apparently because the physician registry in those potential controls was outside of West Cumbria; in seven of those exclusions the fathers had worked at Sellafield. For the included young people, the fathers' preconception exposure to ionizing radiation was then compared for the cases and controls. (Exposure was known with more accuracy than in most cases because radiation badges were worn at Sellafield; for fathers working elsewhere exposure was assumed to be 0.)

The odds ratios were high. In one calculation, using local controls, the odds on death from leukemia for children whose fathers had received at least 100 milliSieverts of ionizing radiation was more than eight times the odds of such death for children of fathers who had no such exposure. Despite this, the court rejected the Gardner hypothesis principally, but not solely, because it lacked biological plausibility. The role of such factors in assessing causation is discussed in the next section.

In the ephedra case described at p. 65, plaintiffs' experts relied, among other things, on a case–control study. In the study, investigators had enrolled 702 patients admitted to hospitals who had suffered a stroke, and knew what diet pills, if any, they had taken 2 weeks before their strokes. As soon as a patient was enrolled, investigators searched by random telephoning for two matching controls – two people with the same gender, age, and geography as the enrolled patient, but who did not have a stroke – who could answer the same question as to diet pills for the same 2-week period used for the enrolled patient. From this data, the investigators computed an adjusted retrospective odds ratio of more than five, meaning that the odds on ephedra use given a stroke were more than five times greater than the odds on ephedra use given no stroke. Hence in the prospective sense – the point of interest in the lawsuit – the odds on a stroke given ephedra use were more than five times greater than such odds when there was no ephedra use.[5]

Biases and Confounding

Epidemiology is not a razor-sharp tool for detecting causal relations. Case–control studies are particularly prone to bias because of the difficulty in identifying controls who match the cases in everything except the disease. Cohort studies tend to be better because it is easier to define groups differing only in exposure (prior to disease development), but even in such studies a finding of association may be produced by biases, errors, confounding, or chance, and not by causal connections. Here are examples of some of the problems.

---

[5] *In re* Ephedra Products Liability Lit., 393 F. Supp. 2d 181 (S.D.N.Y. 2005).

Confounding–We saw in Chapter 2, at p. 37, that there was an association between egg consumption and ischemic heart disease in data for 40 countries. But this association may only mean that a *confounding* factor – such as meat consumption – is associated with both. Problems of confounders are particularly serious for ecologic studies, such as that one, in which associations based on averages (in this case, countries) are used to infer causation for individuals.

Selection bias – The cases or controls may be selected in a way that makes them significantly different from the populations they are supposed to represent. In the Dalkon Shield case, the study was conducted after extensive and negative publicity, which caused most women to have Shields removed. Those women still wearing the Shield at the time of the study might have been more careless about their health than the control women, and this could have accounted for at least part of their higher odds of PID. On the other hand, it could be argued that many of those who would contract PID from the Shield had already done so before the study recruitment period began, and the exclusion of such persons from the study caused an understatement of the association between the Shield and PID. In either event, the cases may not have reflected the general Shield-wearing population of women they were intended to represent.

In the Sellafield case, the court rejected the Gardner hypothesis as unproved for a variety of reasons. One of those reasons reflects an important general problem in such cases: the excess of leukemia that was investigated was to some extent identified in advance by a television program and prior studies. While neither side claimed that the excess was entirely due to chance, the degree of association between leukemia in children and parental exposure at a nuclear facility would have been increased by the post hoc selection of an area with a known excess of leukemia and from which nuclear workers were drawn. This is sometimes called the Texas sharpshooter fallacy after the tale of a "sharpshooter" who demonstrated his proficiency by shooting at random at a barn and then painting bull's-eyes around the holes.

In the neuroblastoma case, the issue of selection effect was debated by the party experts. Defendant's experts claimed that plaintiffs' expert study was scientifically invalid because of the Texas sharpshooter issue that arose for two reasons. Plaintiffs' expert was led to do the study by lawyers for the plaintiffs who were "prompted specifically by their knowledge of cases of neuroblastoma in the community," and there was no prior documentation of the relationship between exposure to the toxins and neuroblastomas. In reply, plaintiffs' expert acknowledged that such studies had the weakness that "the hypothesis about a source may have been formed, and some of the boundaries of the study area chosen, because of informal knowledge about the number of cases nearby." But in their case, the expert claimed there was prior knowledge that the contaminants caused cancer (including neuroblastomas) and the choice of study area was based on the location and extent of the hazard and not on the spatial pattern of the disease.

Response bias – This bias arises when subjects respond inaccurately to an investigator's questions. In the Sellafield case, the court discounted a study of the effect of preconception X-rays of the fathers on the incidence of leukemia on their subsequently born children because the information was obtained by recall from the mothers, who may have been more likely to recall such X-rays if the child had contracted leukemia than if the child's development had been normal.

Ascertainment – Changes in reported incidence of a disease associated with some environmental or risk factor may in fact be due to changes in the accuracy of diagnosis or classification of the disease. In 1976, after an outbreak of swine flu in Fort Dix, New Jersey, the federal government began a massive vaccination program. After about 2 months, a moratorium was declared after cases of a rare neurological disorder, Guillain–Barré Syndrome (GBS), were reported following vaccination. The federal government agreed to compensate vaccinated persons whose GBS was caused by the vaccination. In one case, the plaintiff, one Louis Manko, contracted GBS some 13 weeks after vaccination. This was more than a month after the program had stopped. Because of the unusually long time between vaccination and acute onset, the government denied that the vaccination had caused Manko's GBS and refused compensation. At the time Manko's GBS had appeared, the relative risk of GBS in vaccinated vs. unvaccinated persons was, by plaintiff's computation, almost four. However, one reason for the elevated relative risk was that there had been a sharp decline in the rate of GBS among *un*vaccinated people after the moratorium; if the pre-moratorium rate had been used the relative risk would have been less than two. The parties agreed that this decline was due to underreporting, since GBS was so rare and difficult to diagnose and after the moratorium attention was no longer focused on the disorder. Plaintiff's expert argued that the higher relative risk was nevertheless correct because there was similar underreporting of GBS for the vaccinated cases. This seems unlikely because compensation was paid in reported cases. Nevertheless, the district court found for Manko on this issue.[6]

Association vs. Causation

Whether it is reasonable to infer causation from association is a major question in epidemiologic studies. In resolving that question, epidemiologists look to certain general guides eponymously called the Bradford-Hill criteria, which are really nothing more mysterious than organized commonsense. They are described below.

Strength of association – The larger the relative risk associated with exposure, the less likely it is to be due to undetected bias or confounding. But even large relative risks may be weak evidence of causation if the number of cases is small so that chance alone could explain the association. To measure sampling uncertainty, confidence intervals for the point estimates of the relative risk are commonly computed.

---

[6] Manko vs. United States, 636 F. Supp. 1419 (W.D. Mo. 1986), *aff'd in part*, 830 F.2d 831 (8th Cir. 1987); D.A. Freedman & P.B. Stark, "The Swine Flu Vaccine and Guillaine-Barré Syndrome: A Case Study in Relative Risk and Specific Causation," 23 Evaluation Review 619 (1999).

Consistency – The more consistent the findings from all relevant epidemiological studies the less likely it is that the association in a particular study is due to some undetected bias. All relevant studies should thus be examined. What is sufficiently relevant can be a matter involving technical judgment. In the Vanderbilt case discussed in Chapter 6, in which pregnant women were given a tracer dose of radioactive iron, plaintiffs' expert argued that the Vanderbilt situation was unique because it involved fetal exposure from an internal emitter, and other fetal studies of prenatal radiation involved external emitters. Defendant's expert countered that whether the emitter was internal or external affected only dose and, based on the external emitter studies, calculated that the estimated fetal dose was far too small to have caused the cancers.

Dose–response relationship – If risk increases with increasing exposure it is more likely that the corresponding association would be causal. This is a strong indication of causality when present, but its absence does not rule out causality. In the epidemiological study of the connection between childhood leukemia and pollution of two town wells in Woburn, Massachusetts, (see pp. 44–46) investigators found an increase in relative risk of leukemia for children in households that took a greater percentage of their water from the contaminated wells.

Biological plausibility – There must be a plausible biological mechanism that accounts for the association. This is sometimes said to be a weaker criterion because a biological scenario can be hypothesized for almost any association, and plausibility is often subjective. In the Sellafield case, the court rejected paternal preconception radiation as a cause of the leukemias on the ground that only about 5% of leukemia could be attributed to inheritance. Recognizing this difficulty, plaintiff proposed a synergy theory: *ppi* produced genetic mutations that only caused leukemia when they were activated by an *X* factor, such as a virus or environmental background radiation. The court rejected this argument, pointing out, among other things, that the theory presupposed a high rate of mutations that was inconsistent with both animal and *in vitro* studies. Moreover, the theory did not explain the absence of any excess of other heritable diseases in the children. To reconcile the excess of leukemia with normal rates of other heritable diseases would require either an exquisitely sensitive leukemia gene or an *X*-factor that was specific to leukemia. While plaintiffs' experts offered some theories to that effect, the court rejected them as speculative.

Temporality – Cause must precede effect. This *sine qua non* for causation is usually satisfied. However, some patterns of temporality may affect the plausibility of the causal inference, as when the time between the exposure and the onset of disease is inconsistent with causality. In the swine flu/GBS case, the government argued that the 13 weeks between the vaccination and acute GBS was too long to be consistent with causation.

Analogy – Similar causes should produce similar effects. Knowledge that a particular chemical causes a disease is some evidence that a related chemical may also cause the same disease. This is related to the plausibility argument, and a similar caution is warranted.

Specificity – If risk is concentrated in a specific subgroup of those exposed, or in a specific subgroup of types of diseases under scrutiny, the association is more likely to be causal. This is a strong indication when present. In the Vanderbilt case, plaintiffs' expert argued there was specificity of outcome because two of the four cancers in the children were sarcomas. Defendants' expert replied that since the four cases were a leukemia, a sinovial sarcoma, a lymphosarcoma, and a liver cancer, the diseases were different and specificity was not present.

Experimental evidence in man – Experimental evidence in humans (i.e., random allocation to exposure or removal of exposure) would constitute the strongest evidence of causation. However, such data usually are not available, except in infamous cases of experiments on people.

As should be evident from the foregoing, the derivation of causation from epidemiologic data is not a simple or mechanical process.

Chapter 10
Combining the Evidence

When Sherlock Holmes and Dr. Watson first met, Holmes astonished Watson by observing that Watson had recently been in Afghanistan. As Holmes later explained, he deduced this by putting together bits of evidence: Watson had been an army doctor because he appeared to be a medical type but had the air of a military man; he had been in the tropics because his face was darkened while his wrists were fair; his left arm was injured; and his haggard looks showed that he had undergone hardship and sickness. Ergo, Watson had been in the Afghan wars.

Thus we are introduced to Holmes's signature method of drawing inferences by combining bits of suggestive data. Deductions in statistics are usually not so heroic, but the idea of combining or pooling information is a central one in statistical description and inference. Specifically, it is very common that data from independent strata are pooled to estimate a single overall number. The advantage of aggregation is that the greater numbers make estimates more reliable and give greater significance to a given disparity. The disadvantage is that the pooled data may be quite misleading. A well-known study showed that at the University of California at Berkeley women applicants for graduate admissions were accepted at a lower rate than men. When the figures were broken down by department, however, it appeared that in most departments the women's acceptance rate was higher than men's. This odd result is called Simpson's paradox. The reason for the reversal in the aggregate figures was that women applied in greater numbers to departments with lower acceptance rates than the departments to which men predominantly applied.

Aggregated and Disaggregated Data

The Berkeley data illustrate a situation in which the comparison of a single acceptance rate for men and women was thoroughly misleading. A more subtle problem arises when a single measure of disparity is appropriate and the question is whether the data from independent strata can be aggregated without biasing the resulting estimate of that number. Interestingly, the sufficient conditions for aggregation without bias depend on the nature of the single measure. For an odds ratio that is the same across strata, the two sufficient conditions for an unbiased estimate are that either

M.O. Finkelstein, *Basic Concepts of Probability and Statistics in the Law*,
DOI 10.1007/b105519_10, © Springer Science+Business Media, LLC 2009

(i) the outcome rates for each group (acceptance rates for men and women in the Berkeley example) are the same across the stratification variable (department in the Berkeley example) *or* (ii) the exposure rate at each outcome level (sex for those accepted and rejected) is constant across the stratification variable. In short, we may aggregate in the Berkeley example to compute a common odds ratio if either the acceptance rates for men and women or the rates of males and females among acceptances and rejections are constant across departments. For an assumed difference in acceptance rates or relative risk of acceptance that is constant across departments, the two sufficient conditions are that either (i) the outcome rate overall (the acceptance rate for men and women combined) or (ii) the exposure rate overall (the proportion of men among applicants) is constant across the stratification variable.

If a common parameter does exist then any weighted average of the parameter estimates in the separate strata will be an unbiased estimate of it. To minimize the variance of the weighted average, the weights are chosen so they are inversely proportional to the variances of the estimates.

The argument for aggregation is that disaggregation deprives data of significance when there is a small but persistent tendency across strata. If blacks are persistently underrepresented, the difference may not be statistically significant in any one stratum but the aggregated data would show significance.

For example, a civic organization sued the mayor of Philadelphia claiming that he discriminated against blacks in the appointment of an Educational Nominating Panel. The panel's function was to submit nominees for the school board to the mayor. The panel had 13 members, all appointed by the mayor, four from the citizenry at large and nine from the heads of certain specified citywide organizations. In the four panels appointed by Mayor Tate between 1965 and 1971 there was one with three blacks, two with two blacks, and one with one black. In this period, the general population of Philadelphia was 60% black and the school-age population was one-third black. Plaintiffs argued that the framers of the city charter had intended that the panel should reflect a cross-section of the community. The case went up to the Supreme Court, which held, among other things, that the 13-member panel was too small for reliable conclusions about discrimination and dismissed the case.[1] This is true if one looks at each panel separately: the underrepresentation of blacks, assuming the source population was one-third black, is not statistically significant for the panel with two or three blacks and is of borderline significance even for the panel with a single black. However, there is a persistent underrepresentation of blacks on the panels, with the result that when the data for the four panels are combined there are 52 selections with only 8 blacks. This underrepresentation is highly statistically significant, assuming a random selection from a population that is one-third black. In this context there does not appear to be a risk of bias from the aggregation.

In considering whether aggregation is appropriate, the courts have used various tests: (1) Are the disaggregated strata so small that statistical tests have no power and disaggregation looks like a ploy to preclude a finding of statistical significance?

---

[1] Philadelphia v. Educational Equality League, 415 U.S. 605 (1974).

(2) Is there a justification for aggregation based on the similarity of strata? (3) Do statistical tests indicate that aggregation would be inappropriate because the data for the separate strata are statistically significantly different?

Coates v. Johnson & Johnson[2] is an example of an employment discrimination case involving race in which an econometric test for aggregation carried the day. Analysis of wage data in such cases frequently involves a multiple regression model (see Chapters 11 and 12) in which employee wages are modeled as the weighted sum of various factors believed to influence them; a dummy variable for race is included in the model to test for its effect, if any, on salary. The weights are the coefficients of the factors that are estimated from the data and the issue is whether the race coefficient is statistically significant. In making his analysis, plaintiffs' expert pooled the data over several years and found that the race coefficient was significant. Defendant's expert argued that the data should have been analyzed separately by year, and if so analyzed the race coefficient was not significant. Defendant's expert performed a test that compared the coefficients of his statistical models estimated from separate years (the Chow test) and found that aggregation was inappropriate, presumably because the coefficients of one or more of the variables in the model fitted to the data for each year separately were statistically significantly different (I say "presumably" because the expert did not actually submit his results to the court, but only gave the bottom line). Plaintiffs' expert did not dispute that result, but used a different (log linear) test, which compared the coefficients for race in successive years, instead of the whole model as in the Chow test, and came up with nonsignificant differences. The district court found that pooling was inappropriate. On appeal, the Seventh Circuit held that, disaggregated by year, the data were not so small as to preclude a finding of statistical significance (the annual data yielded sample sizes that ranged from 509 to 662) and that the district court could reasonably have found that defendant's expert, who had more experience in this type of case, was the more credible expert. Although characterizing the pooling issue as a "close question," the appeals court concluded that the district court could therefore reasonably have found that the Chow test was the proper test and discounted the probative value of the pooled analysis.

The decision in *Coates* is questionable. A Chow test that shows no statistically significant difference in coefficients is good evidence that aggregation does not create serious bias, but a contrary finding does not necessarily mean that a regression based on combined data must be rejected. Since a Chow test is of the statistical significance of the difference in all the coefficients in two models, a statistically significant finding will appear if any pair of coefficients is sufficiently different. There is thus no assurance that a statistically significant Chow test result would relate to the race coefficient that is relevant to the case. Moreover, the race coefficient may have varied from year to year, but still have shown a discriminatory effect in every year, and the Chow test would reflect a statistically significant variation.

---

[2] 756 F.2d 524 (7th Cir. 1985).

In *Coates*, there was no indication in the court's opinion that the employer's practices had changed over the period covered by the aggregated data, if they had changed at all. Given the indeterminacy of the Chow test with regard to the relevant issue in the case, it cannot be assumed from the test results alone that the statistically significant coefficient for race in plaintiffs' regressions was the result of aggregation bias.

In an important subsequent case, *Dukes v. Wal-Mart Stores, Inc.*,[3] the court declined to hold that *Coates* mandated the use of the Chow test to justify aggregating data. Women employees of Wal-Mart Stores brought a class action alleging discrimination by their employer and sought to certify a nationwide class of women plaintiffs. To show the commonality of discrimination around the country, a necessary predicate to certification, plaintiffs' expert performed separate regression analyses for hourly and salaried employees for each of Wal-Mart's 41 regions. The regressions, based on aggregated regional data, showed statistically significant gender-based disparities for all in-store job classifications in all of Wal-Mart's regions.

In defense of disaggregation, defendant argued that since store managers possessed substantial discretion and latitude in setting pay rates, stores acted as distinct units and therefore plaintiffs' expert should have performed a Chow test to demonstrate the appropriateness of combining data to the regional level, citing *Coates*. But the court responded that in *Coates* the Seventh Circuit had emphasized that the case was a close one based on its particular facts and had observed that "pooling data is sometimes not only appropriate but necessary." It held that the proper test of whether workforce statistics should be at the macro (regional) or micro (store or sub-store) level depended largely on the similarity of the employment practices and the interchange of employees. Plaintiffs presented evidence that the stores, while autonomous in certain respects, did not operate in isolation; rather, managerial discretion was exercised "in the context of a strong corporate culture and overarching policies." In addition, plaintiffs showed that store managers, the primary salary decision-makers, experienced frequent relocation among stores. The court concluded that nothing in *Coates* mandated a Chow test in its case, presumably because plaintiffs had shown that stores were sufficiently similar in their wage-setting practices, and certified the class. This result seems right, particularly because the ruling was only on the preliminary matter of class certification, where the test was not whether discrimination had been proved, but whether there was at least some probative evidence to support "a common question as to the existence of a pattern and practice of discrimination at Wal-Mart."

The debate on the thorny issue of aggregation in these cases may have been unnecessary. A composition of the aggregation and disaggregation points of view is to disaggregate when there is risk of bias (e.g., by store in *Dukes*), but then in testing significance to combine the evidence from the various strata. In *Dukes*, this would have involved computing regression models for each store separately and

---

[3] 222 F.R.D. 137 (N.D. Calif. 2004), *aff'd*, 509 F. 3d 1168 (9th Cir. 2007).

then combining the data by computing a weighted average gender coefficient from the store models with weights that were inversely proportional to the variances of the store employee gender coefficients. The variance of the weighted average that results from this choice is one over the sum of the reciprocals of the variances. From this a lower bound of a 95% one-sided confidence interval may be constructed as the weighted average minus 1.645 standard errors, on the assumption of independence across stores, as Wal-Mart claimed.

Another way to combine data from independent strata is to sum up the differences between the observed and expected numbers in each stratum, where the expected number is based on the null hypothesis for the stratum, and then divide that sum of differences by the standard error of the sum. Since the strata are independent by definition, the standard error of the sum is simply the square root of the sum of the variances of those differences. The resulting z-score has a standard normal distribution, given that the null hypothesis is true in each stratum. In the Berkeley example, the null hypothesis would be that the acceptance rates for men and women are the same in each department. Another way of combining the data is to square the disparity between observed and expected values in a stratum, divide by the variance for that stratum, and then sum the terms across the strata. The resulting statistic has a chi-squared distribution, with degrees of freedom equal to the number of strata. The first way is more powerful (i.e., more likely to detect discrimination if it exists) if there is a small persistent effect in one direction (women have lower acceptance rates in all departments); the second way is more powerful if there is a larger effect, but it varies (women are ahead in some departments, behind in others).

Here is another example of combining the data. A publisher was accused of age discrimination in its terminations of copywriters and art directors. On 13 different days over a 2-year period the publisher terminated one or more of these employees, for a total of 15 employees terminated. Were the terminated employees older than the average age of copywriters and art directors? If so, was the difference statistically significant? One could not simply aggregate the data by computing the average age of the terminated employees and comparing it with an average age of the employees at any point in time because the average age of employees declined over time.

To answer these questions, the statistician involved in the case treated each day on which a termination occurred as a separate and independent stratum. Given that a termination occurs on a day the null hypothesis is that the selection was made at random with respect to age from the employees on that day. The observed statistic is the aggregate age of the terminated employees on each such day (usually only one, but on one day, three). On each day, the aggregated age(s) of the terminated employees were the observed data and the expected aggregate age under the null hypothesis was simply the average age of all employees on the day times the number of employees terminated. In 10 out of 15 terminations the ages of the terminated employees exceeded the average age at the time of termination. The difference between observed and expected aggregate ages was not significant for any single day. But the statistician combined the data by summing the difference between the observed and expected aggregate ages on each termination day over the 13 days. The aggregate age of the 15 terminated employees was 653 years compared with an expected aggregate age of 563.4 years, for a total of 89 years in excess of expectation under the hypothesis of random selection. This was about 2.5 standard deviations above expectation, a disparity that was significant at the 1% level.

Meta-analysis

Combining the evidence is a valid and informative procedure when data are collected with similar measurement techniques in each stratum, under similar conditions of observation and study protocol. In cases in which epidemiological evidence is important, there are frequently multiple studies of a similar or related problem. It is tempting in such cases to apply the techniques of combining the evidence to a broader problem: to select, review, summarize, and combine the results from separate published studies on a common scientific issue. Such a "study of studies" is called meta-analysis.

If a group of epidemiologic studies is homogeneous, in the sense of all measuring the same thing and differ only because of sampling variation, a summary relative risk may be computed as a weighted sum of the log relative risks of the individual studies. Logs are taken to make the distribution of the sum more normal. The weights are inversely proportional to the variances of the studies. Taking anti-logs retrieves the summary relative risk with a confidence interval.

Whether Bendectin was a teratogen is an example because many case–control studies were made of this question (on the power of these studies, see p. 10). In the litigation, defendant introduced a summary of ten of these studies. Each study generated a log relative risk of malformation in children whose mothers who had been exposed and unexposed to Bendectin in pregnancy and a log standard error for that relative risk. The studies varied in the numbers of cases and controls and in the definition of malformation. The results also varied, some showing a small increased risk and others a small decreased risk, but none being statistically significant taken separately. Combining these results by weighting the log relative risks of malformation inversely to their squared log standard errors and then taking anti-logs gave a common relative risk of 0.96 with a 95% confidence interval of 0.85–1.09. The point estimate of the relative risk is less than one and is not statistically significant. In addition, combining the study results raised their power. Plaintiffs' expert argued that Bendectin increased the risk of malformations by 20%. But against that alternative hypothesis the power of the combined results of the meta-analysis was 80%. So the lack of significance was not due to a lack of power but to the varying results of the studies. Clearly the case for the null hypothesis that Bendectin does not cause birth defects was significantly advanced by this meta-analysis.

In the Bendectin example the case for meta-analysis is quite strong because the studies are quite similar. But meta-analysis in general remains somewhat controversial because meta-analytic results differ in their interpretability, degree of validity, and the extent to which they can be generalized.

Two different studies rarely measure precisely the same parameter. Differences in study methods and measures, subject populations, time frames, risk factors, analytic techniques, etc., all conspire to make the "common" parameter different in each study. Nor is it always clear that one wants to have the same parameter in each study: Many replications of a study demonstrating the lack of a toxic effect of a drug in non-pregnant women would not be informative for pregnant women.

One must then ask what one is testing or estimating when one combines a heterogeneous group of studies. The answer is some ill-specified average of the estimated parameters from the selected studies. Whatever the average is, it depends very much on the sampling scheme by which the studies were selected for combination. This is a serious problem because so often there is no "scientific" sampling of studies. Obviously, bias in the selection of studies is compounded when one limits the selection to all *published* studies, because of the well-known bias that favors publication of significant over nonsignificant findings. This is the "file-drawer" problem: how to access the unknown number of negative findings buried in researchers' file drawers to achieve a balanced cross-section of findings. Finally, there is an almost inevitable lack of independence across studies. Successive studies by the same researcher on a topic are often subject to the same systematic biases and errors; a recognized authority in a field may determine an entire research program; and in extreme cases, scientific competitors may have hidden agendas that affect what gets studied and reported.

For these reasons, results of meta-analysis will be most generalizable if the protocol includes the following steps: (1) creating a preestablished research plan for including and excluding studies, which includes criteria for the range of patients, range of diagnoses, and range of treatments; (2) making a thorough literature search, including an effort to find unpublished studies; (3) assembling a list of included and excluded studies, and in the latter category giving the reasons for their exclusion; (4) calculating a P-value and a confidence interval for each study; (5) testing whether the studies are homogeneous, i.e., whether the differences among them are consistent with random sampling error and not some systematic factor or unexplained heterogeneity; (6) if a systematic difference is found in subgroups of studies – e.g., cohort vs. case–control studies – making separate analyses of the two groups; (7) if the studies are homogeneous, calculating a summary statistic for all of them together, with a confidence interval for the statistic; (8) calculating statistical power curves for the result against a range of alternative hypotheses; (9) calculating the robustness of the result, namely, how many negative studies would have to exist (presumably unpublished, or in a foreign language not searched) for the observed effect to be neutralized; and (10) making a sensitivity analysis, i.e., eliminating a study or studies that appear to have serious design flaws to measure the effect on the results.

In the massive litigation over silicone breast implants the federal cases were consolidated before Judge Sam C. Pointer, Jr., of the Northern District of Alabama. Judge Pointer appointed a panel of non-party experts to examine the evidence that silicone breast implants caused systemic connective tissue disease ("CTD"). The panel took two years and finally issued a voluminous report. One chapter consisted of a meta-analysis of available epidemiologic studies. The report analyzed the connection between implants and five conditions with established diagnoses (e.g., one was rheumatoid arthritis), plus what the individual study authors regarded as "definite connective tissue disease" even though the diagnoses did not fit into established categories. This last category was designed to allow for some uncertainty and change in diagnoses. The report used as the exposure of interest *any* implant, not merely silicone implants, because information on type of implant frequently was missing or could not be verified. Separate analyses were also made of known silicone implants.

The panel searched published databases and a database of dissertation abstracts for unpublished sources. Studies published only in abstract or letter form were included even though they had not appeared in refereed journals. The search was limited to English and to human subjects. Submissions were also received from a plaintiffs' law firm. The criteria for inclusion were (1) an internal comparison group; (2) available numbers from which the investigators could identify the numbers of implanted women with and without the disease and the numbers of non-implanted women with and without the disease; (3) the exposure variable being the presence or absence of breast implants; and (4) the disease variable being some kind of CTD. For each study, the panel calculated an odds ratio or relative risk and an exact confidence interval.

In producing an overall summary estimate, it was assumed that the studies were homogeneous, i.e., that they constituted repeated measurements of the same odds ratio or relative risk. The authors tested this assumption. Finding heterogeneity, they stratified the studies in an attempt to find homogeneity within strata. The chief available stratification variables were study design (cohort or other), medical record validation of disease (yes or no), and date of data collection (before or after publicity relating to the breast implants). When homogeneity could not be achieved through stratification, the investigators resorted to inspections of individual studies in search of outliers. Studies were removed individually or in pairs to achieve homogeneity among the remaining studies. The final set of studies represented the largest number of studies and subjects that were homogeneous.

One study was much larger than the others and in the weighted sum dominated the estimate. The investigators combined the data with and without this study. The results were essentially the same. One argument for exclusion was that, as noted at p. 101, it was the only study in which reports of disease were not based on or confirmed by medical records.

In many of the studies, the authors had adjusted their estimates for possible confounders. The primary confounders were age, with controls frequently being older than the cases, secular time, because implant frequency and type varied by time period, and length of follow-up, because this sometimes varied between implanted and non-implanted women. In a number of studies some adjustment seemed indicated because the unadjusted odds ratios suggested that breast implants had a protective effect against CTD. When the investigators did a meta-analysis combining the results of 14 studies with adjusted odds ratios, they found no significant association between breast implants and CTD. After the issuance of this report, no new cases asserting claims of CTD were brought and almost all of the plaintiffs in the class actions accepted defendant manufacturers' settlement offers that had been made prior to the report. Combining the evidence through meta-analysis is not easy to do well, but the results can have a powerful impact.

Chapter 11
Regression Models

The ancient Greeks personified what happens in life by three conjoined fates: Clotho, the spinner, who represents the fateful tendencies in life; Lachesis, the measurer, who stands for the accidental within those tendencies; and Atropos, the inexorable, who snips the thread, and is death. Putting aside death, the notion that what happens in life is the product of tendencies and random accident within those tendencies – Clotho and Lachesis – is thus an old idea.[1] Its modern counterpart in statistical parlance is called a regression model.

The worldview of regression models is that measurable forces produce the average outcome, which can be predicted from those underlying causes, but that individual outcomes will vary unpredictably around that average as a result of a host of random factors. The measurable forces are called *explanatory factors* or *independent variables* because they "explain," or "account for," the outcome. The outcome is called the *dependent variable* because its average is explained by the independent variables. The dependent variable is said to be *regressed on* the explanatory factors, which are its *regressors*. The accidental influences are all combined in what is called the *error term* because the explanation is only of the average outcome and the differences between the individual outcomes and the averages are viewed as the indeterminacy or error in the equation. It is a fundamental assumption of the regression model that the error term is uncorrelated with the explanatory factors and that its average value is zero. Note that if the error was correlated with one or more explanatory factors, the correlated part could be accounted for by the explanatory factors and would no longer be part of the error. And if the error term did not average to zero, the average outcome could not be predicted from the explanatory variables, as the model assumes, but would have an unpredictable or random component.

The term "regression" was first given to this type of model by the nineteenth-century English geneticist, Sir Francis Galton, who collected data and observed that children of very tall or very short parents tended to be closer to average height than their parents were; as Galton put it, the average height for such children regressed toward the mean height of all children. It is this characteristic that gives the

---

[1] This interpretation is Freud's in his 1913 essay, *The Theme of the Three Caskets.* Republished in 4 *Collected Papers* 245 (New York: Basic Books, Inc. 1959).

M.O. Finkelstein, *Basic Concepts of Probability and Statistics in the Law,* 127
DOI 10.1007/b105519_11, © Springer Science+Business Media, LLC 2009

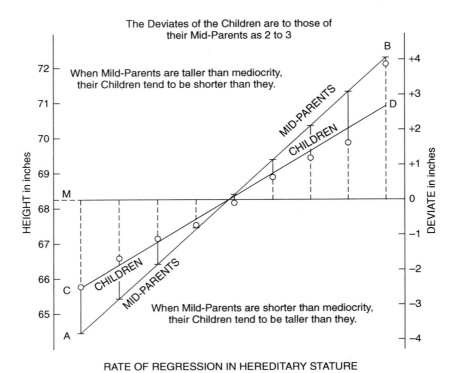

The Deviates of the Children are to those of
their Mid-Parents as 2 to 3

Fig. 11.1 Galton's illustration of regression to the mean

regression model its name. The larger the error term the greater the regression to the
mean. Figure 11.1, Galton's illustration, shows the regression to the mean in his
parent/children data.

Regression models are ubiquitous in economics and social science. In law they
appear in many contexts, but perhaps most widely in employment discrimination,
antitrust, voting, environment, and to some extent in securities cases. We use the
employment discrimination context to illustrate the terms we have been describing. Consider the salaries of employees at a large employer. A very simple salary
regression model would be

$$Y_i = \alpha + \beta X_i + \varepsilon_t$$

where Y_i, the dependent variable, is the salary of the ith employee; X_i is some variable for the ith employee that explains some of the variation in average salary, such
as years employed at the employer (seniority); α is a constant and β is the coefficient of the years-of-seniority variable (α and β are constants in the sense that
they do not vary from employee to employee). Finally, ϵ_i is the error term for the

salary of the ith employee, it being the difference (positive or negative) between the salary of the ith employee and the average salary for employees with that seniority.

Regression models are usually expressed in terms of averages, i.e., in our example, the average salary (denoted $\overline{Y}$) for given values of the explanatory factor, years-of-seniority. Since the mean of the error term is zero, it disappears, and the model is simply

$$\overline{Y} = \alpha + \beta X.$$

In short, the model posits that average salaries for any given level of seniority lie on a straight line, which has intercept α (salary at hire, when seniority is zero) and slope β (average dollar increase per year of seniority).

When there is a single explanatory variable the regression model can usually be represented as a line through a scatterpoint graph of the data. The point of the exercise is to project the value of the dependent variable for a given value or values of the explanatory factor when there is little (or no) data at those values. The model in effect fills in what is weakly covered or missing in the data, and if believed is thought to provide more reliable estimates than the data themselves when the latter are not adequate. It may also (much more questionably) project values for the dependent variable outside of the observed range of the data for the explanatory factors.

Many models in legal proceedings use more than a single explanatory factor and the choice of explanatory variables is a key aspect, and the most debated part, of model building. Suppose employees who reached managerial levels were given an annual bonus of \$1,000 on top of the seniority increase. In that case, if managerial status were not included as an explanatory factor, the extra \$1,000 would be part of the error term for managerial employees. But if, as is likely, managerial status is correlated with seniority, the estimate of b would be increased, in effect attributing to seniority an increase due to the correlated factor of managerial status. In such cases, we say that the regression model is *misspecified* in the sense that an explanatory factor should have been included and was not, and this biased the coefficient of the factor that was included. The remedy for such bias is, as one would expect, to add explanatory factors to the model so that all major influences are accounted for. If managerial status was also a factor, the common practice is to add a "dummy" variable, say X_2, that is 0 when the employee is non-managerial and 1 when the employee is a manager. (This coding is arbitrary; it could equally be expressed as 1 for non-managerial status and 0 otherwise.) With this addition, the model would look like this

$$Y_i = \alpha + \beta_1 X_{1,i} + \beta_2 X_{2,i} + \varepsilon_i,$$

where X_1 is the years-of-seniority variable and X_2 is the dummy variable for managerial status. As before, ε_i is the error term, i.e., the difference between the actual and the estimated salaries after accounting for both seniority and managerial statuses. The average salary of employees $(\overline{Y})$ after accounting for seniority and managerial status would be written as

$$\overline{Y} = a + b_2 X_1 + b_2 X_2,$$

with a and b estimated by a method to be described. If more explanatory variables are needed to model the productivity of employees, these must be added to the equation. The dependent variable, average salary, is modeled as the weighted sum of the productivity factors with the weights given by the coefficients of those factors.

When there are multiple explanatory variables, the point of interest may either be to synthesize their effect to estimate the dependent variable or to analyze the separate effects of individual explanatory factors. As an example of analysis, in a class action brought by women teachers at an academic institution in which they allege discrimination in pay and promotion, the point of interest is whether gender has had an effect on salary apart from other factors that legitimately may influence salary. A statistical analysis of this question may be done in various ways, but the most common practice is to construct a regression model that includes a 1–0 (dummy) variable for gender, the variable taking the value 1 if the teacher is a man and 0 if the teacher is a woman. If the coefficient for that variable is statistically significantly greater than 0, that is taken as evidence that gender was a factor in setting salary in addition to the productivity factors included in the model.

We now illustrate the use of regression for analysis and synthesis with four relatively simpler models that reflect the variety of its uses and some of the objections that have been encountered in litigation.

Four Models

Pay Discrimination in an Agricultural Extension Service

In *Bazemore v. Friday*,[2] the Supreme Court had before it for the first time the use of a regression model in an employment discrimination case. The plaintiffs, black county agricultural extension agents, claimed discrimination in pay. They introduced a regression model with variables for productivity (education, seniority, and title) and a dummy variable for race. The coefficient for race was $394, indicating that after accounting for the productivity factors, blacks earned on an average $394 less than equally qualified whites.

All regression models are simplifications and it is common for those opposing them to claim that some essential explanatory variable was omitted. In *Bazemore*, defendants argued that because the model lacked explanatory variables, such as county contributions to salary, it should not have been admitted in evidence. The Court rejected that argument, holding that "While the omission of variables from a

---

[2]478 US 385 (1986).

regression analysis may render the analysis less probative than it might otherwise be, it can hardly be said, absent some other infirmity, that an analysis which accounts for the major factors must be considered unacceptable as evidence of discrimination.... Normally, failure to include variables will affect the analysis' probativeness, not its admissibility."[3] This holding is a bit ironic because the Court in effect recognized that there *was* a major factor omitted from the model – the amount contributed by the county to agent salaries, which varied from county to county – which the Court analyzed in an ad hoc way to illustrate that such a factor would not account for the disparity in black and white agent salaries. Perhaps more significantly, the Court indicated in a footnote that it was not enough merely to object to a regression on the ground that some factor had been omitted; a party challenging a regression must demonstrate that the omitted factors would change the regression results. "Respondents' strategy at trial was to declare simply that many factors go into making up an individual employee's salary; they made no attempt that we are aware of – statistical or otherwise – to demonstrate that when these factors were properly organized and accounted for there was no significant disparity between the salaries of blacks and whites."[4] This was on point since the defendant clearly had the county contribution data to make such an analysis. But the requirement laid down by the Court cannot be pushed too far because in many cases data are not available for the missing factor or factors – that may be why they are missing. Nevertheless, despite some shakiness, the themes sounded in this case have resonated in much subsequent litigation, as the following example illustrates.

Nitrogen Oxide Emission Reduction

The US Environmental Protection Agency was directed by 1990 amendments to the Clean Air Act to establish a program to reduce nitrogen oxides (NO_x) emissions from coal-fired, electric-utility boilers, a cause of acid rain.[5] The Act contemplated retrofitting boilers with "low NO_x burner technology" (LNB). In Phase I of the program, the Act established for certain types of boilers a maximum emission rate of 0.50 lbs of NO_x per million British thermal units (lb/mmBtu), unless the EPA found that this rate could not be achieved using LNB. In Phase II, EPA was permitted to revise this limit downward if it determined that "more effective low NO_x technology was available." Since LNB had already been installed in connection with Phase I, data were available on the percentage reduction in uncontrolled emissions achieved by the LNB technology to determine whether an emission rate below 0.50 lbs could be achieved with such technology, and if so, what that limit would be. The agency

---

[3] *Id.* at 400. The Court added in a footnote the sensible caution that "there may, of course, be some regressions so incomplete as to be inadmissible as irrelevant; but such was clearly not the case here." *Id.* n. 15.

[4] *Id.* at 404, n. 14.

[5] 42 U.S.C. 7651f (2000).

thus determined to estimate emission rates for retrofitted Phase II boilers and then to set as a new limit the rate that would be achievable by 90% of them.

In making these determinations, the agency did not use a simple average percentage reduction from Phase I units to estimate reductions for Phase II units. The reason was that units with higher uncontrolled emissions had larger percentage reductions with LNB than units with lower uncontrolled emissions, so use of an average figure would understate the reductions achievable by high-end emitters and overstate those of low-end emitters. The EPA believed that the net effect of this would be to understate the effectiveness of the controls and make the total allowed emissions too high.[6]

To make more precise estimates, the EPA used linear regression to model the average percentage reduction in emissions with LNB as a function of the rate of uncontrolled emissions. One model fitted to certain Phase I data (20 units) was $P = 24.9 + 31.1U$, where P is the average percentage reduction in emissions for units with U rate of uncontrolled emissions. The data for this group of Phase I units with the regression line added are shown in Fig. 11.2.

EPA's regression model did not go unchallenged by the utility industry. Several utilities sued, and the case became a landmark in *Appalachian Power Company v. Environmental Protection Agency.*[7] In its challenge, the industry attacked EPA's use

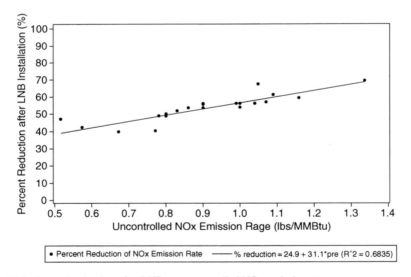

Fig. 11.2 Percent reduction after LNB vs. uncontrolled NOx emission rate

---

[6]EPA, Acid Rain Program; Nitrogen Oxides Emission Reduction Program, Part II, 61 Fed. Reg. 67112, 40 CFR Part 76 (1996).

[7]135 F.3d 791 (D.C. Cir. 1998).

of the regression model. We describe one of its basic objections here and deal with some others later in this chapter.

Appalachian Power argued that EPA's reliance on the regression model was arbitrary and capricious because the model failed to include as explanatory variables operational factors affecting LNB, such as aging and wear of equipment, increased particulate emissions, auxiliary equipment design, and furnace configuration, all of which, Appalachian argued, could have an effect on the level of NO_x emissions. In rejecting this argument, the *Appalachian Power* court cited and followed *Bazemore*. The Court of Appeals adopted the Supreme Court's teaching when it quoted with approval the EPA's response that, "the claim that there are various problems due to aging of equipment that have not yet been encountered is speculative and unsupported." It concluded that "Neither the commenters before the EPA nor Appalachian Power here before us has offered any data to support the assertion that additional factors not accounted for in EPA's regression analysis would have a significant effect on NO_x emissions."[8] The model survived and the EPA won its case.

But models do not always survive missing-variable challenges. In one case, a model used to estimate the effect of an Indiana statute imposing restrictions on abortions was rejected by the court as lacking essential variables.[9] We must, however, defer discussion of that case until after we cover some other topics necessary to understand the model in the next chapter.

Daily Driving Limits for Truckers

In 2005, after a tangled history, the Federal Motor Carrier Safety Administration issued a new rule that raised the daily driving limit for truckers from 10 to 11 hours, among other changes.[10] Data from a study of large-truck fatal accidents showed that the rate of fatigue-caused fatal accidents rose from 4.4% (22 such accidents out of 496 trucks) for hour 10 of driving to 9.6% (9/94) for hour 11 of driving. Thus the rate of fatigue-related fatal accidents more than doubled between 10 and 11 hours. In weighing costs and benefits of the increase, as it was required by statute to do, FMCSA did not accept the conclusion that the rate doubled; in its view the small number of fatigue-related crashes at the 11th hour made the data unreliable. Instead, it fitted a cubic regression model to the data. Each hour of driving was shown as a data point, except (crucially) that accidents involving 13 and more hours were aggregated as a single point at 17 hours (8/32). The curve showed a much smaller increase in fatigue-related accidents at the 11th hour than the doubling shown by the data. See Fig. 11.3.

---

[8] 135 F.3d at 805.

[9] A Woman's Choice–East Side Women's Clinic v. Newman, 305 F.3d 684 (7th Cir. 2002).

[10] The rule reimposed the 11-hour limit that had been struck down by the D.C. Circuit Court of Appeals in a prior case.

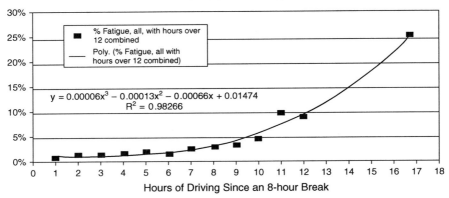

Fig. 11.3 Crash risk as a function of hours of driving

The plaintiffs, Public Citizen among others, challenged the model on the ground that the curve depended crucially on the outlier point at 17 hours and FMCSA had not explained why it had used that number. In a post-hearing letter to the court, FMCSA's counsel explained that 17 was the average number of hours driven post-12. But the court of appeals refused to accept this attempted repair because it was not submitted at the hearing in time for comment, and on this and other grounds again struck down the rule.[11] Arguments over data are also a common ground for contention in regression analysis cases.

Bloc Voting

Here is a more complicated example, involving two regression equations. The Voting Rights Act of 1965, as amended in 1982, forbids any practice that would diminish the opportunity for the members of any racial group to elect representatives of their choice. In *Thornburg v. Gingles*,[12] plaintiffs attacked a redistricting plan that they alleged would dilute their voting power. When the case reached the Supreme Court, it held that to make out a case of vote dilution prohibited by the statute, plaintiffs would have to show, among other things, that (i) the minority is geographically compact (i.e., there is a geographical district in which the minority would be a majority), (ii) the minority is politically cohesive (i.e., it votes substantially as a

---

[11] Owner-Operator Independent Drivers Association Inc. v. Federal Motor Carrier Safety Administration, 2007 WL 2089740 (D.C. Cir. 2007).

[12] 478 U.S. 30 (1986).

bloc for its desired candidate), and (iii) the majority is also a voting bloc (i.e., it generally votes to defeat the minority's candidate).

Since the ballot is secret, plaintiffs in vote-dilution cases have had to use indirect methods of proof to meet the Supreme Court's requirements. One way this has been done is to estimate two regression equations based on data from some prior election. In one equation, the dependent variable is the percentage of registered voters in the district voting for the minority candidate and the explanatory variable is the percentage of minority registered voters in the district. In the other equation, the dependent variable is the percentage of registered voters who vote for any candidate and the explanatory variable is again the percentage of minority registered voters in the district. The two equations are estimated from data for all districts. The regression method described here is called "ecological" because it uses data for groups to make inferences about individuals (i.e., their propensity to bloc vote).

To understand the method, assume that the minority in question is Hispanic. The two relationships are assumed to be two linear regression equations. The first is $\overline{Y}_h = \alpha_h + \beta_h X_h$, where $\overline{Y}_h$ is the average percentage of registered voters in the district who turn out for the Hispanic candidate, α_h is a constant, X_h is the percentage of Hispanic registered voters in the district (based on Hispanic surnames), and β_h is the percentage point increase in the vote for the Hispanic candidate for each percentage point increase in the percentage of Hispanic registered voters in the district. The coefficients of the model are estimated by fitting the equation to the data for all districts in some preceding election. The model tells us that, on average, in a 0% Hispanic district α_t percent of the registered voters (all non-Hispanics) will vote for the Hispanic candidate, while in a 100% Hispanic district, $\alpha_t + \beta_h$ percent of the registered voters (all Hispanics) will vote for the Hispanic candidate. The key premise of the model is that it applies to all districts, both those with low percentages of Hispanic registered voters and those with high percentages, subject only to random error; this is called the constancy assumption.

The voting rates are estimated by a second regression equation, $\overline{Y}_t = \alpha_t + \beta_t X_h$, where $\overline{Y}_t$ is the average percentage of registered voters who vote for any candidate; α_t is the percentage of non-Hispanic registered voters voting for any candidate; X_h is, as before, the Hispanic percentage of registered voters in the district; and β_t is the percentage point change in voting associated with each percentage point change in Hispanic registered voters in the district. In a 0% Hispanic district, on average, α_t percent of the registered voters (all non-Hispanics) would vote, while in a 100% Hispanic district, $\alpha_t + \beta_h$ percent of registered voters (all Hispanic) would vote. (Note that β_h could be positive or negative, corresponding to the fact that voting rates are higher or lower in Hispanic vs. non-Hispanic districts.)

In a 100% Hispanic district, the ratio $\overline{Y}_h / \overline{Y}_t$ is the average proportion of voting Hispanics who voted for the Hispanic candidate; in a 0% Hispanic district the ratio is the proportion of voting non-Hispanics who voted for the Hispanic candidate. Under the constancy assumption these percentages are assumed to apply to all districts. The difference between them is a measure of bloc voting.

In *Garza v. County of Los Angeles*,[13] Hispanic voters in Los Angeles County sought a redistricting to create an Hispanic-majority supervisory district. Plaintiffs' expert estimated the two regression equations using data from a prior election for sheriff. Based on election returns and census tract information from some 6,500 precincts, the first equation was $\overline{Y}_h = 7.4\% + 0.11X_h$. In a hypothetical 0% Hispanic precinct, the second term drops out and the equation tells us that, on average, 7.4% of the registered voters (all of whom are non-Hispanic) would cast their ballots for the Hispanic candidate; under the constancy assumption this figure applies to non-Hispanics in ethnically mixed precincts as well. In a hypothetical 100% Hispanic precinct, the equation tells us that, on average, $7.4\% + 0.11 \times 100\% = 18.4\%$ of the registered voters (all of whom are Hispanic) would cast their ballots for the Hispanic candidate; by the constancy assumption this figure also applies to Hispanics in ethnically mixed precincts. The 11 percentage point difference between turnout for the Hispanic candidate by Hispanics and non-Hispanics is an indication of polarized voting.

The expert's second regression equation was estimated as $\overline{Y}_t = 42.6\% - 0.048X_h$. In a precinct that was 0% Hispanic, on average, 42.6% of the voting-age population would cast ballots, while in a precinct that was 100% Hispanic, on average, $42.6\% - 0.048 \times 100\% = 37.8\%$ would cast ballots. Combining the results from the two equations, the percentage of Hispanic voters who would vote for the Hispanic candidate is $18.4\% / 37.8\% \approx 49\%$, or about half. On the other hand, the percentage of non-Hispanic voters who would vote for the Hispanic candidate is $7.4\% / 42.6\% \approx 17\%$. The difference is obviously substantial, although not as great as those appearing in some later cases. Despite a strong challenge to the regression model, the court in *Garza* found that there was bloc voting and directed a redistricting to provide for a supervisory district with a Hispanic majority.

Use of regression models in vote-dilution cases should be compared with a more direct approach: simply looking at support rates for the Hispanic candidates in districts that are largely Hispanic and non-Hispanic. This is known as *homogeneous precinct analysis*, or *extreme precinct analysis*; it has been used as an alternative measure in some cases. However, it suffers from the drawback that there may be no such precincts or that voter behavior in such precincts may not be broadly representative of voter behavior in more ethnically mixed precincts.

Indeed, the main point of attack on the ecological regression model is its assumption that Hispanics and non-Hispanics in all types of districts have the same propensity for bloc voting, apart from random error. This may not be true if socioeconomic factors associated with the percentage of Hispanic registered voters in a district affect the strength of attachment to a Hispanic candidate. For example, voters in poor, largely Hispanic precincts may attach greater important to the ethnicity of the candidate than Hispanic voters in more affluent and ethnically mixed precincts. Nevertheless, the courts, beginning with the Supreme Court in *Gingles,* have generally

---

[13]918 F.2d 763 (9th Cir.), *cert. denied*, 111 S. Ct. 681 (1991).

accepted ecological regression methods, along with others, as proof of bloc voting in vote-dilution cases.[14]

Estimating the Model

In the usual case, the relationships described by the regression model are not known. In the employment discrimination case, all the employers show us are the salaries and years-of-seniority, education and perhaps other productivity factors relating to the employees, without a description of how those salaries were reached. In scientific pursuits we would say that nature shows us the outcomes (i.e., the salary) but does not tell us which part is the product of systematic influence and which part is error. To derive a regression relation in the usual case, we have to choose a form of regression model, and then estimate the coefficients by fitting the model to the data.

These two aspects of the problem are handled in different ways. The form of model is usually in large part dictated by a combination of convention, the desire for simplicity, the availability of data, and reasonableness. Linear models are a common choice. In our employment example with a seniority variable, the linear model reflects the assumption that each year of seniority is rewarded on an average with the same dollar amount of salary increase. We made this assumption for simplicity, but in fact, in the context of salary, it is neither conventional nor reasonable. The average dollar amount of increase per year of seniority is likely to depend on the level of seniority. In particular, employees nearing retirement would tend to receive smaller increases, reflecting the fact that additional experience beyond a certain point does not increase productivity. Thus regression estimates of salary, using our simple model, would tend to be too high for very senior employees. Clearly a model that makes the average increase depend on the level of seniority would be preferable. How that is usually done will be discussed later. I note that the model might be approximately true for employees within a range of seniority.

The second step – that of estimating the constant term α and the coefficients of the explanatory factors – is essentially mechanical. Each possible set of values generates a set of estimated salaries for the employees. Which set shall we pick? The conventional choice is the set of values that makes the estimated salaries most closely "fit" the actual salaries. The estimated values for α and β are usually denoted by a and b, or sometimes by $\hat{a}$ and $\hat{b}$. Goodness-of-fit is measured in terms of the sum of the squared differences between the actual and the estimated salaries taken over all the employees in the database. The smaller this sum the better the fit. In all but the simplest cases it is necessary to use a computer program to pick the set of values for a and b that minimizes the sum of these squared differences. These values are called the ordinary least squares (OLS) solution to the regression equation. Squared

---

[14]*See, e.g.*, Bone Shirt v. Hazeltine, 336 F. Supp.2d 972 (S.D. 2004)(and cases cited).

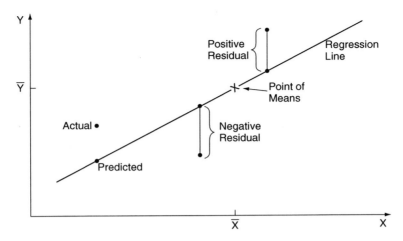

Fig. 11.4 Residuals from a fitted regression model

differences rather than simple differences are used because all squared differences are positive; taking the values for which their sum is minimized gives in most cases a unique solution, whereas there are many possible assignments of values for which the sum of simple differences would be zero, as positive and negative differences cancel out. Minimizing the sum of the absolute differences is a possibility, but lacks certain mathematical advantages of squared differences (Fig. 11.4).

The OLS estimates have the important properties of being both *unbiased* and *consistent*. Unbiasedness means that if we took all possible samples of data from some universe and computed regressions each time, the average of the estimated coefficients would be the true value; there is no systematic error. Consistency means that as the sample size increases the estimated coefficient will converge on its true value.

The differences between the regression estimates of salary and the actual salaries are called *residuals*. The residuals are to the estimated model what the errors are to the true model. The variance of the residuals is used to estimate the variance of the errors.

Measuring Indeterminacy in Models

There are two sources of indeterminacy in regression models. The first is the inherent limitation that the model predicts only averages and individual observations will vary around the average. The second is sampling error due to the fact that in most cases the equation is estimated from a sample of data. Let us consider each of these sources.

Inherent Variability

Two commonly used measures of how well the model fits the data are the squared multiple correlation coefficient and the standard error of regression.

The multiple correlation coefficient, R, is the correlation between the regression estimates and the observed values of the dependent variable. The squared multiple correlation coefficient, denoted R^2, is the proportion of the total variation in the dependent variable that is explained by or accounted for by the regression equation. R^2 ranges between one (the regression estimates explain all variation in the dependant variable) and zero (the regression estimates explain none of the variation). In the salary example this would be the sum of the squared differences between the regression estimate of the salary of each employee and the mean salary of all employees divided by the sum of the squared differences between the actual salary of each employee and the mean salary for all employees. The complement of R^2, which is $1 - R^2$, is proportion of total variation of the dependant variable that is *not* explained by the model, i.e., the sum of the squared error terms divided by the sum of the squared differences between actual salaries and their mean. Figure 11.5 shows these relationships.

A very high value for R^2 is an indication in favor of the model because it explains most of the variation in the dependent variable. There are, however, no hard and fast rules for what constitutes a sufficiently high value of R^2 for one to conclude that the model is valid. Much depends on the context. Generally, in social science pursuits, if R^2 is greater than 50%, the model would be described as a good fit. But in a time-series model, in which the dependent variable is some value in time (e.g., the closing price of a public company's stock from day to day), an R^2 of 90% or more would not be unusual. This is a very good fit, but might only reflect concurrent trends in the dependent and explanatory variables rather than a causal relation. In that case, the model would fail when there was a turning point in the data for one side that

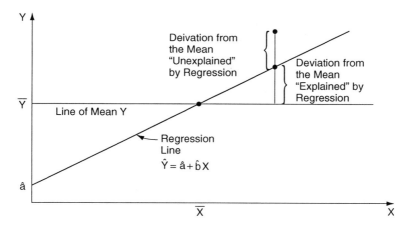

Fig. 11.5 Explained and unexplained deviations from the mean

was not matched by turning points in the data for the other side. Thus, R^2 should be viewed as only one of a number of preliminary tests of a model's validity. On the other hand, if R^2 is low that is a fairly good indicator *against* the model. In the *Appalachian Power* case discussed earlier, the EPA justified the use of a model in which the dependent variable was percent reduction in emissions, which required a second step to calculate the reduced amount of emissions, over a model in which the dependent variable was the reduced amount of emissions because the R^2 of the former was higher (73.1%) than the latter (59.7%).

The other measure of performance is called (confusingly) the *standard error of regression*; it is the square root of the mean squared residual. The larger the standard error of regression the weaker the model in the sense that the data show greater variation that is not explained by the included factors. Some statisticians prefer the standard error of regression to R^2 as a criterion for selecting a regression model from several candidates.

Sampling Error

The second source of indeterminacy arises from the fact that the regression coefficients must be estimated from a sample of data and are thus subject to sampling error. This raises the question whether they are statistically significant. The statistical significance of a coefficient is usually determined by dividing the value of the coefficient by its (estimated) standard error. The resulting ratio is a *t*-statistic (usually called a *t*-value) with degrees of freedom equal to the number of observations less the number of explanatory factors. Coefficients with *t* values less than 2 are usually regarded as nonsignificant and larger ones may also be nonsignificant, depending on the degrees of freedom.

Confidence and Prediction Intervals in Relation to the Regression Estimate

Since the coefficients of the equation will vary due to sampling error, the regression estimates of the dependent variable will have a variance. The positive square root of the variance is the *standard error of the regression estimate*. This, in general, is different from, and will be larger than, the standard error of regression, which reflects only the inherent variability of the regression model, while the standard error of the regression estimate reflects both inherent variability and sampling error. In large samples (e.g., with degrees of freedom greater than 30), a 95% *confidence interval* for the regression estimate is about two standard errors on either side of the regression estimate. The interpretation of this confidence interval is that if we took repeated samples of data with the same values of the explanatory factors and computed regressions from these data, 95% of the confidence intervals would cover the "true" average value of the dependent variable for those values of the explanatory factors. To obtain a *prediction interval* for a single new observation, the confidence interval for the regression estimate has to be widened to allow for the variation in individual observations around the regression estimate, which, as you will recall, is measured by the standard error of regression. The interpretation of a 95% prediction interval is that, in repeated samples with the same values of the explanatory factors, 95% of the computed prediction intervals will cover a new observation of the outcome variable.

Confidence and prediction intervals may be illustrated by a case involving cash and future wheat prices on the Chicago Board of Trade. When these prices rose dramatically during the last few days of trading in May 1963, the Department of Agriculture accused a firm of market manipulation. The firm argued that wheat prices had been artificially depressed and that the dramatic price rise was attributable to normal supply and demand factors. The staff of the Department of Agriculture presented a regression of the cash price for wheat on supply and demand factors. Since the actual average price for May was very close to the regression mean and within the confidence interval, the study was successfully used to rebut the claim that wheat prices had been abnormally depressed. Alternatively, the consistency of wheat prices with supply and demand factors in the last few days of trading might have been tested by comparing those prices with a prediction interval for each day based on the regression model.

Here is another example. In the *Appalachian Power* case previously discussed, Appalachian objected to the use of a regression model to set emission limits because the regression estimates of controlled emissions were subject to error; since many boilers were close to the emission limits any error in the prediction would render those boilers in violation of the limits. In addition, Appalachian pointed out that regression estimates were only averages, and individual boilers could emit more or less than the average. In fact, for any given value of uncontrolled emissions, the controlled emissions of half the retrofitted boilers would be above the regression estimate.

Appalachian's points are not without substance. But the court replied that in any regression analysis the line described by the regression represents the best possible fit to the data, and the fact "that the model does not fit every application perfectly is no criticism; a model is meant to simplify reality to make it tractable." As long as residuals were "within acceptable statistical limits" [whatever that means] the model was not rendered invalid by their presence, "to invalidate a model simply because it does not perfectly fit every data point would be to defeat the purpose of using a model."[15] This response is a bit of a straw man because the issue was not whether the regression had to be perfect, but whether EPA should have made allowance for its uncertainties in setting limits. More persuasively, the court later in its opinion noted that the limits were an annual average so that boilers could over-emit on some days

---

[15] 135 F.3d at 806 (quoting from Chemical Mfrs. Ass'n v. EPA, 870 F.2d 177, 1264 (5th Cir. 1989) (interior quotes omitted).

and under-emit on others and stay within the limits; utilities could also petition the EPA to average across affected units within a company. These are better, if partial, points. To allow for errors in the regression estimate (which would not be cured by averaging) the EPA might appropriately have used the upper end of a confidence interval as a cushion for the uncertainties in the regression estimate. Appalachian's case is weaker for using the upper end of a prediction interval (which would be still higher) because of the provisions for averaging cited by the court.

Models that have smaller confidence and prediction intervals for the dependent variable are to be preferred to models that have greater indeterminacy. For simplicity, consider a model with a single explanatory factor. The size of a confidence interval for such a model depends on the standard error of the coefficient of the explanatory variable (which reflects sampling variability in estimating the model) and the standard error of regression (which reflects the inherent indeterminacy of the model). The interval is not constant, but is smallest at the mean value of the explanatory factor and increases as the explanatory factor moves away from its mean. The reason for this is not that the variance of the constant term a or of the slope coefficient b changes (they don't) but the fact that variation in b has a greater effect on the dependent variable, as one moves away from the mean of the explanatory factor, much like a small change in the angle of a see-saw produces a bigger change in height at the end of a board than nearer the center. At the mean value of the explanatory factor (the pivot of the see-saw) the variability of b contributes nothing to variability of the outcome variable, and the size of the confidence interval is determined solely by the variance of a.

The size of the standard error of a coefficient depends on (1) the standard error of regression, (2) the sample size from which the equation is estimated, (3) the spread of the values of the explanatory factor in the sample data, and (4) where there are multiple explanatory factors, the extent of linear correlation among them. Let us consider each of these in turn.

If the standard error of regression was zero, then the regression equation would predict the dependent variable perfectly and each sample of data with fixed values of the explanatory factors would have the same value of the dependent variable, regardless of sample size. This means that the same regression equation would be estimated from sample to sample, i.e., the standard errors of the coefficients would be zero. As the error variance increases, the influence of the random part of the equation causes the dependent variable to vary from sample to sample and this produces different OLS equations. On the other hand, for any given error variance, increasing the sample size diminishes the variability of the OLS estimates due to sampling.

As for the spread of the explanatory factor, consider the simple case of a regression of Y on X. The OLS line runs through the point defined by the means of Y and X. If the values of the explanatory factor X cluster near the mean, while the Y values are spread apart, even small changes in the Y values can cause large swings in the estimated slope of the line (the b coefficient), as it responds like our see-saw. But if the X values are also spread apart, the line is more stable.

In more complex situations involving multiple explanatory variables, the fourth factor – the extent of linear correlation among the variables – has an important influence on the standard errors. But we defer discussion of that problem to the next chapter, where more complex models involving multiple explanatory models are discussed.

It is important to remember that confidence or prediction intervals for the estimates of the dependent variable reflect sampling variability when the model is estimated from a sample, assuming that the model is correct, and are not an allowance for doubts about the model more generally. If sampling is not appropriate then neither is confidence or prediction intervals.

Before it abolished the death penalty, New Jersey required a proportionality review of death sentences to make sure that the sentence of death was proportionate to the offense, given other capital cases. Regression models were used in which the dependent variable was the probability of a death sentence and the explanatory variables were factors relating to the heinousness of the homicide. In *State v. Loftin*,[16] one of the regression models reported in the opinion (estimated from data for prior penalty trials) yielded a predicted probability of 14% for a death sentence, which was quite low compared with other proportionality review defendants. (Using the same regression model, the court found that the comparable estimated probabilities of a death sentence in the four earlier proportionate review cases were 64, 80, 59, and 65%.) But the court put Loftin's lower figure to one side, commenting that since a confidence interval for the estimate in Loftin's case ran from 0 to 95%, "[that] tells us that we can vest Loftin's numbers with little significance."[17] One may question, however, whether an allowance for sampling error was appropriate in Loftin's case. It was appropriate if there were some true or underlying culpability index that has to be estimated from the available data of prior cases. But if the exercise is simply to compare Loftin's sentence in relation to, or using, sentences in prior cases, then those cases are what they are and there is no reason to view them as a sample.

In December 2007, the New Jersey legislature abolished the death penalty, thereby becoming the first state in 40 years to do so by legislative action.

I now turn to more complex models and their issues.

---

[16] 157 N.J. 253 (1999).

[17] Models based on different databases put Loftin's probability of a death sentence at levels that were higher than other proportionality review defendants; the court did not object to the results of those models on the ground that the confidence interval was too wide.

Chapter 12
Regression Models – Further Considerations

The preceding chapter discussed the simplest models: those in linear form (in one case in cubic form), and without transformations of data. Many models in legal proceedings raise more complex issues than those already discussed and involve the appropriateness and interpretation of explanatory variables, alternative kinds of models, transformations of data, and even systems of equations. Some of these complexities are the subject of this chapter.

Choice of Explanatory Variables

Ideally, the choice of explanatory factors would be determined by some relevant theory that provides a basis for selecting factors and does not depend on the particular available data. In a discrimination case, a labor economist might guide a party in identifying factors that relate to employee productivity. However, there are frequently a number of plausible candidate factors and theory does not always inform the choice among them. When that is the case, the selection is usually made by a computer using a variety of algorithms. In one popular algorithm, known as the forward-selection method, the computer selects as the first variable the one that best explains (i.e., is most highly correlated with) the dependent variable and computes a regression using that single variable. The computer then picks as the next variable the one that best explains the remaining unexplained variation in the dependent variable (i.e., the prediction errors in the model computed with the first variable). This second variable is added to the model, which is then recomputed. The computer proceeds in this way until no more variables can be found that, when added to the equation, will significantly increase its explanatory power. There are variants of this basic idea.

Methods of this class are called "stepwise" and are sometimes criticized as "data-dredging." The reason is that when a regression model is selected and estimated by fitting a model to data the fit may seem very good, but there is usually a deterioration in performance when the model is used to make predictions for other data sets. The greatest deterioration occurs when the regression has been constructed using a forward-selection procedure or some other data-dredging technique. (The extra good fit has been called "overoptimism.") To correct for overoptimism, if sufficient data are available, it is desirable to select and estimate the equation from one data set, but to measure its goodness-of-fit using another set. Use of the forward-selection method has led to the rejection of models by one district court,[1] but generally the main battles lie elsewhere.

Proxy Variables

It frequently occurs that an investigator does not have data for all plausible variables and feels compelled to use proxies for them. This may raise a variety of problems: The proxy may be biased in the sense that it doesn't fairly reflect what it is supposed to represent; its inclusion may change the meaning of other explanatory variables in the equation; or it may bias the coefficient estimates of other variables. Here are some examples of these problems.

A relatively benign example involved EPA's model used in the *Appalachian Power* case discussed in Chapter 11. Appalachian Power objected that the regression model used short-term (52-day) rates of uncontrolled emissions. This was inappropriate, the utility argued, because EPA used this data to generate estimates of *long-term* (annual) rates of emissions and to set long-term limits. The agency responded that it used short-term data as a proxy for long-term emissions for the pragmatic reason that sufficient long-term data were not available. As a further response, the EPA reported that it tested for bias by estimating a model using long-term data for the 18 units for which such data were available. The fit was deemed "acceptable" ($R^2 = 65.3\%$). Applying that model, the agency justified its use of short-term data when it found that the percentage of boilers estimated to achieve a specified emission limit using the long-term data usually varied by less than 2% (and not more than 5%) from the percentage estimated using short-term data. It is rare that the validity of a proxy can be tested as the EPA was able to do in this case.

Here is an example of a biased proxy. In a seminal case some years ago a class action was brought against Republic National Bank for discrimination against blacks and women.[2] In their regression models, plaintiffs used personal characteristic variables and job variables to adjust for differences in groups of employees. Among the personal characteristics were education (highest grade completed) and age (which was used as a proxy for the explanatory variable of general labor-market experience). Fitted to the data for several thousand employees, the model showed that the coefficients for sex and race were statistically significant and adverse to women and blacks. On this result, plaintiffs rested their case. But the court rejected

---

[1] See, e.g., McCleskey v. Zant, 580 F. Supp. 338 (N.D. Ga. 1984), *rev'd on other grounds sub nom.*, McCleskey v. Kemp, 753 F.2d 877 (11th Cir. 1985), *aff'd*, 481 U.S. 279 (1987). This case is discussed at p. 163.

[2] Vuyanich v. Republic Natl. Bank, 505 F. Supp. 224 (N.D. Texas 1980), *vacated*, 723 F.2d 1195 (5th Cir. 1984), *cert. denied*, 469 U.S. 1073 (1984).

age as a proxy for general labor-market experience, finding that it was biased and inaccurate with respect to women because they tended to be out of the labor force and not in school for longer periods than men.

Even if a proxy is unbiased – in the sense that men and women who are equally productive would have the same proxy value – the fact that the proxy is not perfectly reflective of productivity, but only correlated with it, can bias the coefficients of the model. To see this, imagine a nondiscriminatory employer who pays men more than women because men are more productive than women. In a discrimination suit the women produce a regression model in which salary is regressed on an evaluation score (bestowed on employees by an independent company) as a proxy for productivity, and sex. Taking an extreme case, suppose the proxy is essentially uncorrelated with productivity (it's a rotten proxy, but equally uninformative for men and women). Then the inclusion of the proxy will fail to adjust for the difference in productivity between men and women, and the difference in salary would be attributed to sex, i.e., it would be reflected in the sex coefficient, leading to a false conclusion that the employer was discriminatory. When the case is less extreme, i.e., the proxy is correlated but not perfectly correlated with productivity, we say that the proxy underadjusts for productivity; this has the effect of biasing the sex coefficient and may create the appearance of discrimination even if the employer is innocent.[3]

An issue of underadjustment bias was raised by experts for the defense in a case in which women doctors at a medical school claimed discrimination in pay and promotion.[4] The difference in mean salaries for men and women was approximately $3,500, but men had greater experience, education, and tenure at the institution. After adjustment for these proxies, the largest sex coefficient was $3,145 and was statistically significant. But defendant's experts argued that the existence of a statistically significant sex coefficient was not unambiguous evidence of discrimination because it could be the result of underadjustment bias if the institution was nondiscriminatory. The district court thought that the experts were claiming that if the proxies were more accurate (i.e., included more elements of productivity), male faculty would be even more productive than indicated because the unmeasured elements of productivity would also favor men, as the measured elements did. The court

---

[3] When imperfect unbiased *causal* proxies are used – such as years of experience – the problem of underadjustment bias does not arise. This is because the definition of an unbiased causal proxy is one in which men and women with the same value of the proxy will be equally productive, on average, and so a nondiscriminatory employer would pay them equally, on average, and there would be no sex effect. However, an evaluation score is an example of a *reflective* proxy and it is such proxies that may create underadjustment bias. This follows because an unbiased reflective proxy is one in which men and women with the same productivity would have equal proxy scores, on average, but when this is true it will generally be the case, surprisingly, that men and women having equal proxy scores will not be equally productive. Thus the sex coefficient in a regression that uses reflective proxies may be a consequence, to some extent, of underadjustment bias rather than discrimination. For a further discussion of this counterintuitive point see Michael O. Finkelstein & Bruce Levin, *Statistics for Lawyers*, § 14.5 (2d ed. 2001).

[4] Sobel v. Yeshiva University, 839 F.2d 18 (2d Cir. 1988).

of appeals rejected that assumption, and held that "the simple fact of imperfection, without more, does not establish that plaintiffs' model suffered from underadjustment, even though men scored higher on the proxies." But, as we have seen, that can be the result for a nondiscriminatory employer if the proxies are reflective. It does not, however, follow that the underadjustment would necessarily account fully for a statistically significant sex coefficient.

The inclusion of a proxy may also change the meaning of other explanatory variables in the equation, so caution in interpretation is indicated. Here is an example.

Public schools in the United States are usually supported with a combination of local taxes and state funds. In some states, poorer school districts have sued the state claiming that it provides too little support for local school districts to meet the state's constitutional requirement to provide a sound basic education. The argument is that student test scores – as measured by performance on standardized tests – tend to be higher in more affluent districts that exact special levies and spend more on education. The question is whether this is a direct result of the amount spent per pupil, or other correlated social factors not related to expenditure.

In one such case, the Seattle school district sued the State of Washington and the State introduced an education production model to rebut the inference that school variables were important to the levels of scores.[5] The State's model was a regression equation in which the dependent variable was a standardized test score and the explanatory factors were *background* variables (such as parents' educational level and family income), *peer group* variables (such as percentage of minority students and social class composition of students), and *school resource variables* (such as class size, teacher qualifications, and spending per pupil). The relevant question was whether, in such a model, the coefficients for school resource variables were large enough to be practically and statistically significant.

The State included prior test score (district average, same grade, four years earlier) as a proxy variable to account for peer group factors not included in the model that might affect performance. The result was that the coefficients for school resource variables were very small and nonsignificant, which supported defendant's position that spending on schools, within the range encountered in the state, was not a factor in educational outcomes, at least as measured by scores on the standardized tests. However, by including prior test score in the same grade 4 years earlier, the meaning of the equation was changed from prediction of standardized test scores from district to district to prediction of *changes* in standardized scores over the preceding four years from district to district. This is quite a different story. To see this, suppose that school resource factors were important in determining scores, but there had been no significant change in such factors (or others) in the past four years or, as a result, in the level of scores from district to district. In that case the current scores would be essentially the same as the prior scores in each district (subject to random error). In the regression equation the old scores would almost fully explain the current scores and the coefficients of school resource factors and other explanatory

---

[5] Seattle School District v. State of Washington, 90 Wash.2d 476 (1978).

factors in the regression would be small and statistically nonsignificant. But from that result one cannot rightly infer that school resource factors were unimportant to the level of scores.

Tainted Variables

Parties opposing a regression analysis in a discrimination case have raised a variety of objections, e.g.: (1) certain explanatory factors are potentially tainted by discrimination; (2) omission of important wage-explaining variables or inclusion of variables that may not in fact have been wage-explaining; (3) incorrect aggregation of different groups in the same regression; and (4) inclusion of observations from a time when discrimination was not illegal. Here is an example of how the courts have dealt with the tainted variable objection.

A number of actions have been brought against academic institutions for alleged discrimination against women faculty members. Typically, both sides introduce regression models, plaintiffs to demonstrate that after adjusting for legitimate productivity factors, there is residual discrimination reflected in a statistically significant coefficient for gender. Plaintiffs' regressions commonly include variables for education, work experience, publications, and tenure at the school, among others. The defendant institution objects that plaintiffs' models do not capture all of the factors that go into academic productivity, such as quality of teaching. To incorporate those factors, defendant's regressions include academic rank as a proxy for them. When academic rank is included, the gender coefficient frequently becomes statistically nonsignificant. However, plaintiffs object that by including rank the coefficient for gender describes only the discrimination against female faculty *within* a given rank; this does not tell the whole story because there is also discrimination in the awarding of rank. There has been considerable litigation over this issue.

The problem was explored at some length in *Coates v. Johnson & Johnson*,[6] a class action claiming discrimination in the discharge of minority employees. The employer used prior disciplinary actions as an explanatory variable in its regression analysis; its expert found that the number of disciplinary actions taken against an employee in the 12 months prior to discharge was one of the strongest predictors of discharge. But plaintiffs argued that the imposition of discipline was discriminatory in that blacks received formal written warnings for rule violations while whites were given only oral counseling (which apparently left no record). Plaintiffs and defendant introduced anecdotal evidence to support their claims, but there was no statistical evidence and defendant's expert conceded that none was possible. In the absence of substantial evidence, the outcome turned on the burden of proof. The court of appeals reviewed the cases and concluded that once defendant offered a regression with an allegedly biased factor, the "plaintiff must bear the burden of persuading the factfinder that the factor is biased." Since the evi-

---

[6] 756 F.2d 524 (7th Cir. 1985).

dence on this issue was "sparse and conflicting," the court of appeals held that it could not say that the district court "clearly erred in finding that the evidence failed to show that defendant engaged in a pattern or practice of discrimination in issuing disciplines." Consequently, the district court could properly admit defendants' regression study, with the employees' discipline record included as an explanatory variable.

The court distinguished a leading earlier case, *James v. Stockham Valves & Fittings Co.*,[7] which reached an opposite conclusion. In *Stockham Valves*, the employer claimed that a disparity in average hourly wages between blacks and whites could be explained by "productivity factors," among them the variables "skill level" and "merit rating." Plaintiffs objected that the variables were defined in a way that included discrimination and the court agreed. "Skill level" was derived from the employee's job class, but because plaintiffs' claimed that they were excluded from job classes with high ratings, the court found that "[a] regression analysis defining 'skill level' in that way thus may confirm the existence of employment discrimination practices that result in higher earnings for whites." Similarly, the court noted that "merit rating" was based on the subjective evaluation of defendant's supervisors who were overwhelmingly white. The *Stockham Valves* court thus concluded that defendant's regression including these factors did not rebut plaintiffs' prima facie case.

The *Coates* court distinguished *Stockham Valves* on the ground that in *Stockham Valves* there was statistical evidence that "clearly showed a substantial disparity between the representation of blacks and whites in the higher job classes" and showed that "the average merit rating given to blacks was significantly less than that given to whites." According to the court, there was thus support for plaintiffs' allegation that these factors reflected discrimination. By contrast there was no such evidence in *Coates*. But *query*, whether this is the right ground for the distinction. Any explanatory variable that reduces the coefficient of the dummy variable for group status must be correlated with that group-status variable. So such a correlation is not *necessarily* proof that the variable is tainted by discrimination. The question is whether the statistical and non-statistical evidence, viewed as a whole, is a sufficient indicator of discrimination, *vel non*, to shift the burden of proof on the question of taint.

Aggregation Issues

The general pattern is this: plaintiffs' experts like to aggregate data because it gives their regression results greater statistical significance, which they claim is justified; defendants' experts like to disaggregate because it lowers significance, which they say is justified because of aggregation bias. I discussed this issue in Chapter 10.

---

[7] 559 F.2d 310 (5th Cir. 1977), *cert. denied*, 434 U.S. 1034 (1978).

Forms of Models

A variety of regression models has found its way into cases. Taking our discrimination case example, most commonly a dummy variable for gender, say X_3, is added to the model in addition to factors for employee productivity. This can be coded 1 when the employee is a man and 0 when the employee is a woman (or vice versa, the coding is arbitrary and, *mutatis mutandis*, does not affect the results). Then, if the coefficient b_3 is positive, it represents the average dollar amount by which the male salary exceeds the female salary for employees who have the same levels of seniority and managerial status. Conversely, if negative, the coefficient represents the average dollar amount by which the male salary is less than the female salary for such employees. Such a difference, if it is large enough to be statistically significant, generally is taken as evidence of discrimination.

The other kind of model used in discrimination cases does not have a gender variable, but instead has two separate equations – one for men and the other for women. The men's equation is estimated from the men's data and similarly for the women. The coefficients for the explanatory variables are then compared. If, for example, the coefficient for age is larger for men than for women, and the difference is statistically significant, that is an indication that the employer rewards a year of service more highly for men than for women. The advantage of this type of model is that it does not assume a particular form of discrimination, but allows the data to indicate where the discrimination lies. The disadvantage is that more data are required to estimate two equations.

Forms for Variables

A variety of forms may be used for the dependent variable and for whatever explanatory variables are chosen. I give three common examples here.

Logarithms

In many kinds of economic models it is common to express the data for the dependent variable in logarithms. For example, the logarithm of salary may be used in the discrimination model we have been discussing instead of the actual salary. In that case, the coefficient of gender would be interpreted as the approximate percentage difference between men's and women's average salaries instead of the dollar difference. This is probably an improvement because a discriminatory employer would be unlikely to have a fixed dollar difference at all levels of salary; a percentage disparity would seem more likely. Logarithms also have other desirable technical features.

Quadratic Terms

Another alternative form is to add a quadratic term of one or more of the explanatory variables to the equation. Referring again to the salary model, it is common to add a quadratic term for seniority to a salary regression model. The contribution of seniority to salary then has two components: the original bX plus the new b^*X^2. Usually, the estimated coefficient of b^* is negative. Since X^2 increases faster than X, the effect of the negative squared term is to decrease the effect of seniority as X increases, which allows the model to reflect the fact that seniority matters less as it increases.

The appropriateness of a model with a quadratic term came under consideration in the *Appalachian Power* case discussed in Chapter 11. The EPA compared two models. In Model #1 (the one-step approach) the controlled emission rate (C) was regressed on the uncontrolled emission rate (U), the model being $C = a + bU$, as the power companies suggested. In Model #2 there were two steps. In the first step the percentage emission rate reduction (P) was regressed on U, the model being $P = a' + b' U$; in the second step, C was calculated from $C = (U)(1-P/100)$. The agency picked the two-step approach because the model used to calculate P in the two-step model fit the data much better than the model used to calculate C in the one-step model.[8] The reason for this is that the one-step model assumed that controlled emissions rose as rapidly as the uncontrolled emissions, whereas Model #2 allowed for the fact, indicated by the data, that the controlled emissions rose more slowly than the uncontrolled emissions because percentage reductions in emissions were higher when uncontrolled emissions were higher.

The power companies made two related objections to the two-step model. First, they argued that the regression model was of the wrong form, because it used percentage reduction in emissions as the dependent variable, which was described as a "related, but secondary parameter," whereas the quantity of direct interest was the amount of controlled emissions. Second, they argued that the model led to manifestly absurd results since, if the uncontrolled emissions were high enough, the percentage reduction in emissions estimated from the model would rise to 100%! The observation that for extreme values the model yields untenable results is sometimes made in attacks on regression models.

The EPA responded that "the statistically verifiable fit of a regression model is only assured within the range of the data actually used in constructing the model." The level of uncontrolled emissions needed to drive the percentage reduction to 100% was 2.574 lbs per mmBTU, which was far beyond this range. For uncontrolled emissions beyond the range (there was only one such Phase II unit), the EPA put a cap on things: It assumed that the percentage emission reduction would be no greater than the percentage reduction estimated from the model for the unit with the highest uncontrolled emission rate in the database. This unit had a rate of 1.410 lbs per mmBTU.

Making predictions only within the range of the data used to construct the model is generally a reasonable approach, and the fact that beyond that range the model produces perverse results does not necessarily justify the conclusion that it is incor-

---

[8] The correlation between the actual and estimated values of controlled emissions, R^2, is 73.1% for Model #2 compared with 59.7% for Model #1.

rect within that range. In this case, however, there are problems with the appropriateness of the model within the range of the data. As the agency itself pointed out, its two-step model could be equivalently expressed as a one-step model in which C was regressed on U in the quadratic form $C = a + b_1 U + b_2 U^2$. When this model is fitted to the Phase I data, b_2, the coefficient of the quadratic term, is negative, producing estimates of C that grow more slowly as U increases and finally start to decline as U passes a certain (inflection) point. Clearly the model is not valid beyond that point (even if the percentage reduction is not 100%), so there are problems with the appropriateness of the model not just at the extreme value at which the percentage reduction is 100%, but at the inflection point, which is at a much lower level of emissions (1.287 lbs per mmBTU). Since the highest rate of uncontrolled emissions in the data used to construct the model (1.410 lbs per mmBTU) was above the inflection rate, the model overstated the reductions in emissions achievable with LNB at that level of controlled emissions. Exclusion of those points from the data used to construct the model doesn't change the estimated coefficients very much, but it should not have been used, as the EPA said it did, as the cap to set the maximum percentage reduction achievable by Phase II boilers with uncontrolled emissions above that level. The lower inflection point probably should have been used instead. So in assessing the appropriateness of regression estimates one has to go further than merely looking to see whether the values used to generate them are within the range of the data used to construct the model.

Interactive Variables

A third example is an *interactive variable*, which is a product of two variables. Returning to our salary model, if the employer is discriminatory, he may treat gender at the managerial level differently from gender at a lower level. To allow for this, one might include in the model an interaction variable, $b_4 X_2 X_3$, where X_2, as before, is coded 1 for managers and 0 for nonmanagers, and X_3 is coded 1 for a man and 0 for a women. With that coding, the interaction variable would equal 1 only when both Xs equaled 1, i.e., for male managers, and for all other employees would equal 0. The addition of this interactive term changes the meaning of the other coefficients because the effect of being a manager is now assessed separately for men and women. Nonmanagerial women are the base group because for them both X_2 and X_3 are zero. For men and women with equal seniority, the coefficient b_2 now becomes the average increase in salary for managerial over nonmanagerial *women*; and the coefficient b_3 becomes the average increase in salary for nonmanagerial men over nonmanagerial women. The coefficient b_4 reflects the average increase in salary for managerial men over nonmanagerial women. If we wanted to compare managerial men with managerial women, the measure of the average difference in salary would be $b_4 - b_2$. If the variable for gender is also interacted with X_1, years of experience, the coefficient of the interactive variable would be the average added salary per year of experience for nonmanagerial men over nonmanagerial women.

This is what we could learn from estimating separate equations for men and women and comparing the coefficients for the explanatory variables. In general, if sex is interacted with each of the other explanatory variables, the result is equivalent to having two equations, one for men and the other for women.

Consider the following example. A government plant near Cincinnati, Ohio, refined uranium trioxide powder and machined uranium metal for use in nuclear reactors. On December 10, 1984, the contractor announced that over the past 3 months the plant had released over 400 lb of uranium trioxide powder into the atmosphere. Home owners in the vicinity sued for personal injury (primarily fear of cancer) and loss of home value.[9] To support a claim for property damage for houses within 5 miles of the plant, an expert for the plaintiffs presented a regression model based on about 1,000 house sales both before and after the announcement. The dependent variable was house value and the explanatory variables were chosen to reflect housing characteristics and economic conditions that would affect house values. The key regressors were dummy variables defined as follows: *Pre-1985 In*, coded 1 for sales within 5 miles of the plant prior to 1985, otherwise 0; *Pre-1985 Out*, coded 1 for sales beyond 5 miles prior to 1985, otherwise 0; and similarly for *Post-1984 In*. The reference group was *Post-1984 Out*, coded 0 on the three other indicators.

The regression coefficients were *Pre-1985 In*: –$1,621; *Pre-1985 Out*: $550; and *Post*-1984 *In*: –$6,094. The expert took the –$6,094 coefficient to be an appropriate measure of damage for houses inside the 5-mile radius. But this coefficient represented the post-announcement loss in value for those houses within the radius compared with the pre-announcement value of houses outside the radius – the reference group. This was clearly wrong because there was also a negative premium for being inside the radius before the announcement; by the regression model it was $(-\$1,621) - (\$550) = -\$2,171$. The appropriate measure of damage thus had to compare the *change* in the negative premium before and after the announcement. This was $\{(Pre-1985\ In) - (Pre-1985\ Out)\}$, the negative premium for being inside before the announcement, minus $\{(Post-1984\ In) - (Post\ 1984\ Out)\}$, the negative premium for being inside after the announcement. The result is that the negative premium for being inside increased from –$2,171 to –$6,094, an increase of –$3,923. Although still substantial damage per house, plaintiffs' case was weakened because the interactive coefficient was not statistically significant. After a mock trial before a jury by agreement of the parties to assist in evaluating the merits of the claims, the case was settled for a substantial sum, based mostly on compensation for fear of cancer caused to the home owners, with property damage a secondary consideration.

Here is another example of regression with interactive variables. Indiana enacted a statute making a woman's informed consent a condition to an abortion and requiring a waiting period of 24 hours after the information was received. By also requir-

---

[9] In re Fernald Litigation, Master File No. C-1-85-0149 (SAS). The case is discussed in Statistics for Lawyers, § 14.2.1.

ing that the relevant information be supplied by the doctor who would perform the procedure and "in the presence of the pregnant woman" – rather than by printed brochure, telephone, or web site – the statute obliged the woman to make two trips to the hospital or clinic. This raised the cost (financial and mental) of an abortion, with its principal impact on younger and poorer women in areas without a local clinic that provided abortions.

Plaintiffs sued to enjoin enforcement of the two-trip statute on the ground that it imposed an "undue burden" on a woman's right to an abortion. To prove its case, plaintiff relied, inter alia, on a study of the impact of a similar Mississippi statute, which became effective in August 1992.[10] The number of abortions declined after the statute was passed, but to what extent was that decline due to the statute versus a secular trend of declining abortions? To separate these possible effects, the study included a regression model in which the dependent variable was the natural logarithm of the monthly rate of abortions (number of abortions per thousand females aged 15–44) over 72 months, beginning in August 1989 and ending in July 1994; the study period thus included pre- and post-statute abortions. The abortion data were from three states: Mississippi, Georgia, and South Carolina. The latter two states were controls: they had abortion laws that were similar to Mississippi's but did not require the information to be delivered in person, and thus did not require two trips. The explanatory variables included a dummy variable for the two-trip law (law = 1 for abortions after the statute's effective date; 0 for earlier abortions) and dummy variables for the control states (Mississippi was the reference state; for Mississippi residents the dummy variables for the other two states were both 0). The model also included a trend variable (taking on values 1,..., 72 for successive months) and a trend-squared variable to allow for curvilinear trend. To compare the shift in abortion rates after the law in Mississippi with the pattern in the control states, the model included interactive law variables [law (date) x Georgia] and [law (date) x South Carolina] and interactive trend variables [trend x Georgia] and [trend x South Carolina]. The model included 11 dummy variables, one for each month (less the reference month) to capture seasonal trends. Since each of the explanatory variables (except for the seasonal dummies) was interacted by state, the regression was approximately equivalent to having separate regressions for each state.

The effect of the statute was measured by the difference between the coefficient for the law variable for Mississippi and the coefficient for the interactive law variables for the control states. The difference was –0.126 for Mississippi vs. South Carolina and –0.101 for Mississippi vs. Georgia. Since the dependent variable is the natural logarithm of the abortion rate, these coefficients are proportionate changes: The rate of abortions declined about 12% more in Mississippi than in South Carolina after the law, and about 10% more than in Georgia. The study also showed that after

---

[10] T. Joyce, et al., "The Impact of Mississippi's Mandatory Delay Law on Abortions and Births," 278 Journal of the American Medical Association 653 (1997).

the statute's effective date, late-term and out-of-state abortions increased substantially more for Mississippi residents than for residents of the control states.

This study and other evidence led the district court to find that the two-trip statute imposed an undue burden on a woman's right to an abortion; it enjoined the requirement that the information be delivered in the presence of the pregnant woman. On appeal, however, the Seventh Circuit found that differences between Indiana and Mississippi made the Mississippi study a questionable basis for predicting what the effect of the two-trip provision would be in Indiana. In the court's view the regression should have included factors such as urbanization, income, average distance to an abortion clinic, and average price of abortion. By not including such variables the regression assumed what was to be proved, namely, that Indiana was like Mississippi on those factors. According to the court, this shortcoming could have been remedied either by including additional variables in the regression or by gathering data from other states to test whether (and, if so, how) state-specific characteristics affected the results. Since neither was done, the court concluded that the evidence was insufficient and reversed the district court.[11]

Measuring Uncertainty in Models

The discussion in the preceding chapter is applicable here, but with an added point. In more complex situations involving multiple explanatory variables, the fourth factor affecting the standard errors of the coefficients of the explanatory variables – the extent of linear correlation among them – has an important influence on the standard errors. Data exhibiting such correlations are said to be *collinear*; when one factor is nearly a linear function of several other factors, the data are said to be *multicollinear*, a not-uncommon occurrence in multiple regression models. Multicollinearity does not bias the estimates of regression coefficients, but enlarges their standard errors, making the OLS estimates less reliable. In a seemingly reasonable model an explanatory factor with a perverse coefficient (i.e., one that is contrary in sign to what would be expected) may be due to sampling error when the explanatory factor is highly correlated with another factor. The reason for this instability is easy to see. When two factors are highly correlated, assessment of their separate effects depends on the few cases in which they do not move in tandem. In such cases, the usual calculation shows enlarged standard errors, reflecting the smallness of the effective sample size. In effect, the model is overspecified in the sense that more variables are included than are necessary or justified by the data. When one of these coefficients is the subject of importance, a remedy is to eliminate the redundant variable or variables.

---

[11] A Woman's Choice – East Side Women's Clinic v. Newman, 305 F.3d 684 (7th Cir. 2002).

Assumptions of the Model

The usual calculations of confidence and prediction intervals and standard errors of the coefficients rest on four key assumptions about the errors. If these are not satisfied the OLS equation may be biased or inconsistent or the usual computations of sampling variability may misstate the statistical significance and reliability of the results. Let us consider these critical assumptions and the consequences of their failure in some detail.

First Assumption: The Errors Have Zero Mean

The first assumption is that the expected value of the error term is zero for each set of values of the explanatory factors. In that case the regression estimates are unbiased; that is, if repeated samples are taken from the modeled data and regression equations estimated, the overall average of the estimated coefficients will be their true values. When this assumption is not true, the estimates will be biased and if the bias does not approach zero as sample size increases, the estimates are inconsistent. The assumption is violated whenever a correlation exists between the error term and an explanatory factor. One reason this occurs is that some important explanatory factor has been omitted and that factor is correlated with factors in the equation. In the seniority example already given, when there is a bonus for managerial status, but that factor has been omitted from the equation, the expected value of the error term is zero in the early years when no one is a manager, but becomes systematically positive in later years as the model predictions become systematically too low. There is thus a correlation between the explanatory factor of seniority and the error term and that correlation is associated with bias, as we have seen, in the coefficient for years of seniority.

Another reason for a correlation between an explanatory factor and an error term is a "feedback" relation between the dependent and explanatory variable, i.e., the dependent variable also has a causal impact on the explanatory variable. For example, in an antitrust case involving claims of a conspiracy to set prices of cardboard containers, plaintiff introduced a regression equation to prove what the price of containers would have been in the absence of the conspiracy; one of the explanatory variables was the price of containerboard from which the containers were made. But plaintiff's expert had to acknowledge that there was also a reverse relationship: if the price of containers declined because of reduced demand, the price of containerboard would also decline. That feedback relation would mean that when the price of containers was below the regression estimate, i.e., the error term was negative, the price of containerboard would be depressed, creating a correlation between the error term and containerboard price. However, the OLS equation estimated from the data would attribute that correlation to the explanatory factor by increasing its coefficient. This would bias the equation in the sense that it would cause it to overstate the direct effect of the price of containerboard on container prices, and consequently to overstate the effect on container prices of the conspiracy to raise containerboard prices. To avoid such bias when there is a reverse relationship, economists estimate separate supply and demand equations, among other methods.

Second Assumption: The Errors Are Independent

The second assumption is that the errors are independent across observations. Independence means that knowing the error in one observation tells us nothing about the errors in other observations. Errors that are not independent are said to be correlated. If the data have a natural sequence, and correlations appear between errors in the sequence, they are said to be autocorrelated or serially correlated.

Autocorrelation appears most frequently in time series data. It may arise from failure to include an explanatory variable that relates to time or to "stickiness" in the dependent variable. In a price equation based on monthly prices, changes in cost or demand factors might affect prices only after several months, so that regression estimates reflecting changes immediately would be too high or too low for a few months; the error term would be autocorrelated.

Regression estimates are unbiased in the presence of autocorrelation as long as the errors, autocorrelated or not, have an expected value of zero. However, when there is positive autocorrelation, the usual calculation seriously overestimates precision because the variability of residuals underestimates the true error variance. To illustrate, suppose that price data are autocorrelated as against time so that the observations are either first consistently above and then consistently below the true regression estimate, or vice versa (first below, then above). In repeated samples, the over- and underestimates would balance so that the regression value would indeed be the average. However, when the regression model is estimated by OLS from a single sample, the effect of the OLS procedure is to drive the regression line through the observed points and this makes the residuals of the estimated equation smaller than the errors of the true model. Figure 12.1 shows the situation in which the data are first above and then below the true regression line; in that case the slope of the OLS line is less than the true line.

However, if the situation were reversed, and the data were first below and then above the line of the true model, the slope of the OLS line would be greater than the true line. Since either situation can occur, the true line is right on average, but the variation in slope of the OLS lines is greater than would be indicated from either one alone. The equation is unbiased because on average it is right, but the sampling error of the coefficients is greater than that indicated by the usual calculation.

When the equation includes a lagged dependent variable, the existence of serial correlation in the error term may even result in biased estimates. The error at time t is of course correlated with the dependent variable at t. If there is serial correlation in the error term, the error at t is correlated with the error at $t-1$ and, since the error at $t-1$ is correlated with the dependent variable at $t-1$, it follows that the error at t is correlated with the dependent variable at $t-1$. If the lagged dependent variable is

Fig. 12.1 Effect of
autocorrelation

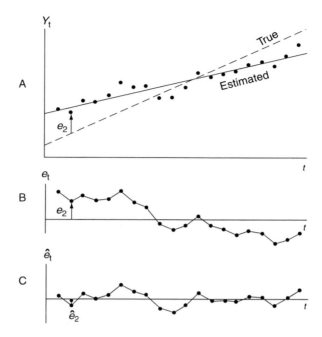

included as an explanatory factor in the regression model, the error at t no longer has zero mean and the correlation between the lagged dependent variable and the error term biases the estimated coefficient of the lagged dependent variable as it "picks up" the correlated part of the error term.

Third Assumption: The Errors Have Constant Variance

The third assumption is that the errors have constant variance about the regression value, irrespective of the values of the explanatory factors. If this condition is met, the data are said to be homoscedastic; if not, they are heteroscedastic (from the Greek word for "scatter"). See Fig. 12.2.

There are two consequences of heteroscedasticity. One is that the usual measures of precision are unreliable. When there is fluctuation in the variance of error terms, the usual calculation of precision uses, in effect, an average variance based on the particular data of the construction sample. If another sample of data concentrated its predictors in a domain of greater variance in the error term, the estimated precision from the first sample would be overstated.

This difficulty may seem esoteric, but it is by no means uncommon, particularly when variables in the regression equation are so structured that the entire scale of the equation (and hence the residuals) increases systematically. This often occurs with salary data since the entire scale of the equation increases with increasing salary.

Son's height

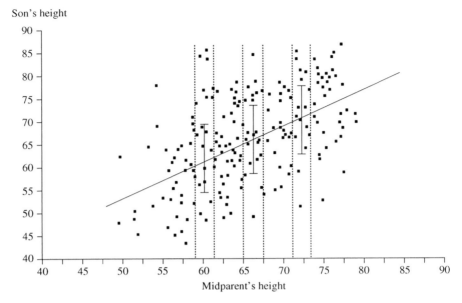

Fig. 12.2 Simulated parents-and-sons data illustrating constant error variance

The other consequence of heteroscedasticity is that the OLS estimates are no longer the most efficient. If some observations have larger errors than others, those observations with large random effects contain less reliable information than observations with small effects and should receive less weight in the estimation. OLS regression, which gives the same weight to all observations, fails to take this heterogeneity into account and, as a result, may have a larger variance for its estimates than other methods.

There are two kinds of remedies for heteroscedasticity. One is *weighted least squares* which, like OLS, estimates coefficients by minimizing a sum of squared deviations from fitted values, but where each deviation is weighted inversely to the variance of the errors. Another type of remedy is to transform the data using one of a number of *variance stabilizing transformations* to reduce or eliminate the heteroscedasticity. For example, salary as the dependent variable in a wage regression equation is frequently transformed into logarithms, in part because the increase in the scale of the equation with increasing salary is much smaller with logarithms than in the original form.

Fourth Assumption: The Errors Are Normally Distributed

The fourth assumption is that the errors are normally distributed in repeated sampling for any given values of the explanatory factors. A violation of this assumption means that the usual computations of precision are no longer reliable. The key is the correctness of the model. If all important influences on the dependent variable are accounted for by the model, what is left are many small influences whose sum is reflected in the error term. The central limit theorem then assures us that, to a good approximation, the error term is normally distributed. In addition, moderate departures from normality do not seriously affect the accuracy of the estimates of statistical significance.

Validating the Model

Whenever the reliability of the estimates is important, it is essential to test for, or at least carefully consider, the correctness of the assumptions of the standard model I have discussed. Since the assumptions concern the errors, and these are unknown, one set of tests involves looking at the residuals.

The OLS protocol forces residuals to have the characteristics of the expected values of the errors in the true model. One characteristic is that the positive and negative residuals balance out, so their algebraic sum is always exactly zero. In the true model, the average sum of errors over all different samples of data (the expected value) is zero, but due to sampling variation the sum of the errors for the particular sample used to estimate the equation will not necessarily be zero.

A second characteristic is that residuals have exactly zero correlation with explanatory factors. (If there were a nonzero correlation between the residuals and an explanatory factor, the fit of the equation could be improved simply by increasing the coefficient for that factor.) In the true model, errors have zero correlation with the explanatory factors. As before, though, this represents an expectation over the theoretical universe of data; due to sampling variation, some sample correlation between the true errors and explanatory factors can exist within the data from which the equation is estimated.

Since the residuals must have these characteristics irrespective of the model, their existence is not evidence that the model is correct. The fact that the residuals sum to zero does not show that the model is unbiased and the absence of correlation between residuals and explanatory factors does not indicate that the model is correctly specified. However, proponents of models can and should look to residuals to pinpoint any outliers, and for indications of nonlinearity, serial correlation, heteroscedasticity, and nonnormality. I discuss each of these subjects below.

The general method for analyzing residuals is to inspect computer-generated plots of residuals against other quantities to detect violations of the assumptions and to compute certain statistics. Residuals are generally plotted in standardized form, i.e., in terms of the number of standard deviations from the mean, which is zero, against standardized values of either the predicted dependent variable or an explanatory variable. The convention is to plot the residuals on the vertical axis and the dependent or explanatory variable on the horizontal axis.

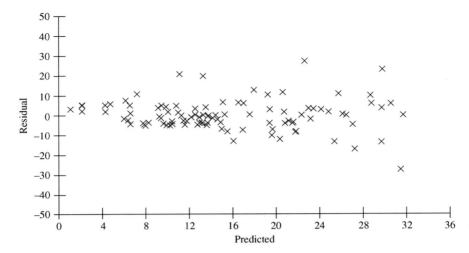

Fig. 12.3   A plot of residuals

Residuals are plotted against predicted values of the dependent variable to see if the data are homoscedastic. A symptom of heteroscedasticity is a residual plot that steadily fans out or narrows in with increasing predicted values. See Fig. 12.3 for an example of residuals evidencing moderate heteroscedasticity.

If the error variance is constant, the cloud of points should look like a horizontal football: the spread is greatest at the center of the plot and smallest at the ends. The reason is that in order to avoid large residuals that would prevent minimization of the residual sum of squares, the regression tends to stick more closely to those values of the dependent variable that correspond to influential extreme values of the explanatory factors than to the less influential middle values. Residual plots of this type will also reveal outliers, that is, points that depart markedly from the pattern of the rest of the data and that may have a large, and possibly distorting, effect on the regression. The solution to the problem of outliers is not simply to eliminate them automatically, but to subject them to detailed examination, possibly leading to their elimination if their accuracy is questionable. Finally, residual plots of this type can be used to test the residuals for autocorrelation. This can be done from visual inspection of the plot, supplemented by computation of a test statistic, such as the Durban-Watson statistic.

Residuals are also plotted against each explanatory factor in the equation. While the OLS equation ensures a zero linear correlation between the residuals and each explanatory factor, there may be a systematic nonlinear relation between them, which would indicate that the model was misspecified by having the wrong functional form (like failing to include a quadratic term) for that factor.

Logistic Regression

In many situations, the outcome of interest is simply whether a case falls into one category or another. When there are only two outcome categories, the parameter of primary interest is the probability or the odds that the outcome falls into one category (a positive response) rather than the other.

The usual linear models should be avoided in this context because they allow the predicted outcome probability to assume a value outside the range 0–1. In addition, OLS is inefficient because the variance of the error term is heteroscedastic: it depends on the probability of a positive (or negative) outcome and the variance of that estimate is smallest when the probability is nearest 0 or 1 and greatest when it is near 0.5.

A standard approach in this situation is to express the outcome odds in a logarithmic unit or "logit," which is the logarithm of the odds on a positive response, i.e., the logarithm of the ratio of the probability of a positive response to the probability of a negative response. The logit is the dependent variable in a regression equation; as is frequently the case, the logit is assumed to have a linear relation to the explanatory factors included in the equation.

One of the most important examples of logistic regression in a law case is *McCleskey v. Kemp.*[12] *McCleskey* was a major challenge to the administration of Georgia's death penalty statute on the ground, inter alia, that a sentence of death was more often imposed in white-victim than black-victim cases. Under the direction of Professor David Baldus, extensive data were collected on the defendant, the crime, aggravating and mitigating factors, the evidence, and the race of the defendant and victim in Georgia homicide cases. From these data, among a variety of analyses, Baldus and his colleagues estimated a number of logistic regression equations. The dependent variable was the log odds of a death sentence, which was regressed on a large number of factors relating to the aggravation of the homicide and dummy variables for race of the defendant and race of the victim. For one group of data (100 cases), in the middle range of aggravation, the estimated race-of-victim coefficient was 1.56, which means that the log odds of a death sentence were increased by adding 1.56 when a white victim was involved. Taking anti-logs, the coefficient becomes an odds multiplier: The odds of a death penalty in white victim cases are $e^{1.56} = 4.76$ times greater than when a nonwhite victim was involved. Note that this odds multiplier is constant regardless of the levels of the other explanatory factors, but its importance depends on the levels of such factors and the estimated odds of a death sentence. Translating odds into probabilities, the model tells us that if the probability of a death sentence was 10% when there was a black victim, it becomes 34.6% when there is a white victim; if it was 50% for a black victim it becomes 82.6% for a white victim.

Ordinary least squares cannot be used to estimate a logistic model because the equation is not linear. An alternative technique is maximum likelihood, which

---

[12] 481 U.S. 279 (1987).

calculates the set of coefficients that maximizes the likelihood of the data. The standard errors of the explanatory factors' coefficients have the same interpretation as they do in ordinary least squares regression, except that for significance testing the normal distribution is used instead of the t-distribution, because the estimates are valid only for large samples.

Goodness-of-fit is not estimated by R^2, but by the proportion of correct predictions of the model. The convention adopted is to predict the more probable outcome (50+%) on the basis of the model and then to calculate the rates of false predictions. In the data we are describing (100 cases), in the middle range of aggravation, the model made 35 correct predictions of a death sentence and 37 correct predictions of no death sentence; the overall rate of correct predictions was thus 72%. Since goodness-of-fit is measured from the same data that were used to construct the model, these results have an element of overoptimism. To tease out the overoptimism, we split the data into two groups, a construction sample and a test sample (50 cases in each group). We then estimate the model from the construction group and then apply it, first to the construction group and then to the test group. This yields the result that the rate of correct predictions falls from 78% in the construction sample to 66% in the test sample. For a more detailed evaluation, these measures of accuracy may be applied to positive and negative predictions separately. The positive predictive value, PPV, of the model in the test sample (correctly predicting a death sentence) was 88%, but the negative predictive value, NPV (correctly predicting no death sentence), was 55%.[13]

In its landmark decision, a narrowly divided Supreme Court upheld McCleskey's conviction and capital sentence. Justice Powell, writing for five justices, assumed that the model was statistically valid (although he repeated without comment the objections of the district court, which had rejected the model), but held that, given the necessity for discretion in the criminal process and the social interest in laws against murder, McCleskey had to prove by exceptionally clear evidence that the decisionmakers in *his* case had acted with intent to discriminate; the statistical pattern was not sufficient for that purpose. The four dissenters pointed out that McCleskey's case was in the middle range of aggravation, for which the statistics showed that 20 out of every 34 defendants convicted of killing whites and sentenced to death would not have received the death sentence had their victims been black. It was therefore more likely than not that McCleskey would not have received the death sentence had his victim been black. His sentence thus depended on an arbitrary racial factor in violation of the Eighth and Fourteenth Amendments. Their view did not prevail, and McCleskey, his appeals exhausted, was then executed.

After his retirement from the Court, Justice Powell told an interviewer that one of the two actions he most regretted while on the bench was his vote in *McCleskey*.

---

[13] The PPV and NPV of the model are referred to as its sensitivity and specificity, respectively.

Table of Cases

(References are to pages where cited.)

Index

Breinigsville, PA USA
01 April 2010
235343BV00002B/13/P

9 780387 875002